SpringerBriefs in Mathematics

SpringerBriefs present concise summaries of cutting-edge research and practical applications across a wide spectrum of fields. Featuring compact volumes of 50 to 125 pages, the series covers a range of content from professional to academic. Briefs are characterized by fast, global electronic dissemination, standard publishing contracts, standardized manuscript preparation and formatting guidelines, and expedited production schedules.

Typical topics might include:

- A timely report of state-of-the art techniques
- A bridge between new research results, as published in journal articles, and a contextual literature review
- A snapshot of a hot or emerging topic
- An in-depth case study
- A presentation of core concepts that students must understand in order to make independent contributions

SpringerBriefs in Mathematics showcases expositions in all areas of mathematics and applied mathematics. Manuscripts presenting new results or a single new result in a classical field, new field, or an emerging topic, applications, or bridges between new results and already published works, are encouraged. The series is intended for mathematicians and applied mathematicians. All works are peer-reviewed to meet the highest standards of scientific literature.

Titles from this series are indexed by Scopus, Web of Science, Mathematical Reviews, and zbMATH.

Daniel Hernández-Hernández •
Joshué Heli Ricalde-Guerrero

Optimal Control in Random Environments

Pontryagin Maximum Principle with Environment-dependent Jumps and Mean-Field Games with Common Poissonian Noise

Springer

Daniel Hernández-Hernández
Probability and Statistics
Centro de Investigación en Matemáticas
Guanajuato, Guanajuato, Mexico

Joshué Heli Ricalde-Guerrero
Mathematics
ETH Zurich
Zurich, Switzerland

ISSN 2191-8198 ISSN 2191-8201 (electronic)
SpringerBriefs in Mathematics
ISBN 978-3-032-25640-9 ISBN 978-3-032-25641-6 (eBook)
https://doi.org/10.1007/978-3-032-25641-6

Mathematics Subject Classification: 60G55, 49J50, 91G15, 60K37, 91A16, 93E03

This Springer imprint is published by the registered company Springer Nature Switzerland AG
The registered company address is: Gewerbestrasse 11, 6330 Cham, Switzerland

Preface

This text is intended to serve as a reference to present updated results on the theory of stochastic optimal control, when state dynamics is a forward process with two different classes of noise. One of them is the standard Brownian motion, while the second is a marked Poisson process with random intensity. One of the main novelties in this text is the inclusion of a random environment, which influences the behavior of the stochastic differential equation (SDE) through its coefficients, as well as the random intensity of the noise with jumps. This random intensity can have an exogenous or endogenous character, depending on the problem at hand, and can represent, in the first case, an extra randomness included in the system, describing a particular status of the environment under study.

The initial motivation for analyzing these types of stochastic processes stems from the theory of optimal mean-field control, where this influence comes from the average position of a finite number of evolving particles (which induces an endogenous random environment). However, after developing the initial results, we realized that modeling an exogenous random environment also had an interesting and meaningful interpretation for many applications. We have therefore worked to extend this idea to include these classes of noise, thus completing a broad set of models to which the obtained results are applicable.

Based on the theory of backward stochastic differential equations, it is well known that these represent a very useful tool for studying optimal control problems, particularly those in which a Markovian structure is not assumed in the model. The intuition behind the dynamic programming principle can be recovered together with some version of the verification theorem. Moreover, when the stochastic version of the Pontryagin maximum principle is satisfied, the value function corresponds to the solution of a backward stochastic differential equation (BSDE) and the adjoint process is well adapted within this theory. This result is successfully extended to optimal control problems evolving in a random environment, providing first some motivation from the calculus of variations and the optimality conditions induced by the Euler-Lagrange equation.

This four-chapter text opens with preliminary material that introduces the deterministic Pontryagin maximum principle, illustrated by its connection with calculus of variatios and an example. This initial chapter also covers necessary concepts on point processes, with a particular focus on marked Poisson processes. Key findings regarding the stochastic calculus of these processes are also reviewed. We then continue with the study of forward–backward stochastic differential equations (FBSDEs) within the framework of a dynamic random environment evolving in the set of probability measures. The existence and uniqueness of a strong solution for such systems is established, illustrating its relevance through several examples, including environment-dependent Markov chains and linear FBSDEs with jumps. This result requires a particular configuration of the probability space, as it is necessary to separate the noise sources.

The exposition of the maximum principle in Chap. 3 analyzes its stochastic version. For controlled processes in a random environment, we provide precise sufficient and necessary conditions to ensure that the backward component—corresponding to the adjoint process—is well-defined and a solution exists. The linear-quadratic model serves to illustrate and motivate the main results. As an application, we also present the portfolio selection problem with regime switching.

In the last chapter, we addressed the existence of a strong solution for SDEs when the random environment is endogenous; for example, when it is based on the conditional law with respect to the noise source of the jumps or, more specifically, on the empirical measure process of the system components. Motivated by the evolution of a finite set of particles, we analyze the propagation of chaos of this system, extending recent results from the theory of mean-field games to the specific framework analyzed throughout this work for models involving a random environment.

The Research Center for Mathematics (CIMAT) is celebrating its 45th anniversary, and we join in their celebration with this text. Professor Wendell H. Fleming's influence during the conceptualization and development of this text was always a motivation for the first author to complete it.

The second author would also like to thank Professor Dylan Possamaï, whose friendship and continuous support were essential to the completion of this project, as well as the students of the course *Topics in Stochastic Optimal Control* (ETH Zürich, Autumn 2025), whose helpful feedback on earlier versions of the notes improved the quality of the present manuscript.

Guanajuato, Mexico
Zurich, Switzerland
March, 2026

Daniel Hernández-Hernández
Joshué Helí Ricalde-Guerrero

Contents

Notation

PMP	Pontryagin's Maximum Principle
SMP	Stochastic Maximum Principle
SDE	Stochastic Differential Equation
FBSDE	Forward–Backward Stochastic Differential Equation
MFG	Mean-Field Game

s.t.	such that
w.r.t.	with respect to
l.h.s., r.h.s.	left-hand side, right-hand side
càdlàg	right continuous with left limits
càglàd	left continuous with right limits

$\mathbb{R}^d$	Set of d-dimensional real vectors
$\mathbb{R}^d_+$	Set of d-dimensional non-negative vectors
$\mathbb{R}^{d\times n}$	Set of $d \times n$ real matrices
$(\mathrm{X}, d_{\mathrm{X}})$	Polish space with metric d_{X}

Id_d	Identity matrix of size $d \times d$
$A^{(i,j)}$	Entry (i, j) of matrix A
$A^{(i,\cdot)}$	i-th row of A
$A^{(\cdot,j)}$	j-th column of A
$A^{\top}$	Transpose of A
$\mathrm{diag}\,(a)$	Diagonal matrix with diagonal a
$(A \cdot B)_{\mathrm{Fr}}$	Frobenius inner product
$\|A\|_{\mathrm{Fr}}$	Frobenius norm

$x^{(i)}$	i-th component of x
$x^{(-i)}$	Vector x without the i-th component
δx	Difference $x - x'$
$(x \cdot y)_G$	G-weighted inner product

$(x \cdot y)$	Euclidean inner product
$\|x\|$	Euclidean norm
c_G	Proportionality constant in $(x \cdot Gx)_G = c_G \|x\|^2$

Let $f : I \subseteq \mathbb{R}^d \to \mathrm{X}$ be a k times continuously differentiable function.

$\mathrm{D} f$	Jacobian of f
$\mathrm{D}^2 f$	Hessian of f
$\mathrm{D}^k f$	k-th derivative of f

$\mathrm{C}(I; \mathrm{X})$	Continuous functions
$\mathrm{C}^k(I; \mathrm{X})$	k-times continuously differentiable functions
$\mathrm{C}^k_b(I; \mathrm{X})$	Bounded derivatives up to order k
$\mathrm{D}(I; \mathrm{X})$	càdlàg functions

$\sigma(\mathrm{A})$	σ-algebra generated by sets A
$\sigma(f)$	σ-algebra generated by functions f
$\mathrm{B}(\mathrm{X})$	Borel σ-algebra on X
$\mathrm{D}([0, T]; \mathrm{X})$	Borel sets in Skorohod topology
$\mathrm{M}(\mathrm{X})$	Finite Borel measures
$\mathrm{M}^*_c(\mathrm{X})$	Counting measures
$\mathrm{P}(\mathrm{X})$	Probability measures
$\mathrm{P}_p(\mathrm{X})$	Measures with finite p-moment
$\Pi(\nu^1, \nu^2)$	Transport plans
$\mathrm{W}_p(\nu^1, \nu^2)$	Wasserstein distance

$(\Omega, \mathrm{F}, \mathbb{P})$	Probability space
$\mathbb{F}^{\mathrm{nat}, M}$	Natural filtration of M
$\mathbb{F}^M$	Completed filtration
$\mathbb{F}' \hookrightarrow \mathbb{F}$	Immersed filtrations
$\tilde{\mathbb{F}}$	Predictable σ-algebra
$\| \cdot \|_*$	Supremum norm
$\mathrm{S}^2(\mathrm{X})$	Adapted càdlàg processes
H^2	Square-integrable martingales
$(M \cdot M')_{\mathrm{H}^2}$	H^2 inner product
$\|M\|_{\mathrm{H}^2}$	H^2 norm
$[M, M']$	Quadratic covariation
$\langle M, M' \rangle$	Predictable covariation
$H^\perp$	Orthogonal martingales
$\mathbb{H}^2(M)$	Space of integrands
$\mathrm{L}^2(M)$	Space of stochastic integrals

Let $\Psi : \Omega \times [0, T] \times \mathrm{P}_2(\mathbb{R}^d) \times \mathbb{R}^d \to \mathbb{R}^n$ be given function.

$\Psi_t(\nu, x)(\omega)$	Value of Ψ at (t, ν, x)
$\delta\Psi_t(\nu, x)$	Difference in (ν, x)

$\delta\Psi_t^\nu(x)$	Difference in x
$\delta\Psi_t^\nu(x^{(i)}, x'^{(i)} \vert x^{(-i)})$	Componentwise difference

Chapter 1
Foundations

Abstract This chapter surveys the core analytical tools underlying the optimal control and stochastic frameworks developed in the book. We first review Pontryagin's Maximum Principle in the deterministic setting, emphasizing its variational structure and the associated adjoint system governing necessary optimality conditions. We then introduce essential elements of stochastic analysis and point process theory, with a focus on integration with respect to Poisson random measures and Lévy–Itô processes.

1.1 Introduction

In this chapter, we provide an overview of the mathematical tools and techniques that will be used throughout the book. From the perspective of optimal control, we start recalling the classical deterministic formulation of *Pontryagin's Maximum Principle* (PMP), a variational method that yields necessary conditions for optimality. On the other hand, from the perspective of probability theory, we introduce the fundamental notions of stochastic analysis and point process theory required in subsequent chapters. Particular emphasis is placed on stochastic integration with respect to integer-valued (Poisson) random measures and on Lévy–Itô processes, which form the analytical backbone for modeling jump dynamics.

The material presented here is largely standard. However, our aim with this chapter is to provide a concise and self-contained reference that offers readers who may not be familiar with some of the topics at hand a coherent introduction to the concepts and results shown later.

1.2 Maximum Principle

The PMP approach for solving optimal control problems dates back to the 1950s, with the work from Boltyanski, Gamkrelidze and Pontryagin [17]. In its original

D. Hernández-Hernández, J. H. Ricalde-Guerrero, *Optimal Control in Random Environments*, SpringerBriefs in Mathematics,
https://doi.org/10.1007/978-3-032-25641-6_1

formulation, the problem consisted in finding a control function that minimizes the transition time of a non-autonomous system between two given states. This is known in the literature as an *optimal time control problem*; however, not long afterward, the principle was generalized to include controls over a more general space while minimizing with respect to an integral functional [16, 36, 63].

Consider a control space A, which is typically a closed bounded set in a Euclidean space. Using our current notation, this translates as finding a control, represented as a measurable function

$$\alpha : [0, T] \to A, \tag{1.1}$$

such that, for a given initial state $X_0 \in \mathbb{R}^d$ and smooth functions $b, f : \mathbb{R}^d \times A \to \mathbb{R}$ and $g : \mathbb{R}^d \to \mathbb{R}$, the controlled system described by the ordinary differential equation

$$\mathrm{d}X_t^\alpha = b(X_t^\alpha, \alpha_t)\,\mathrm{d}t, \qquad 0 \le t \le T, \tag{1.2}$$

minimizes the value of the objective functional $J(\alpha)$ defined in terms of the running and terminal cost functions as:

$$J(\alpha) := \int_0^T f(X_t^\alpha, \alpha_t)\,\mathrm{d}t + g(X_T^\alpha). \tag{1.3}$$

Briefly, PMP consists of finding a characterization of an optimal control α (that is, an optimal control for which the value $J(\alpha)$ is minimized) in terms of the variational (in)equalities satisfied by J when an instantaneous deviation from the optimal control α occurs. The key condition from which PMP receives its name refers to the inequality satisfied by the variation of the objective functional with its conjugate in the phase (state) space, and which can be interpreted geometrically as a supporting hyperplane in an extended phase space (state of the system plus the "running value" of the objective functional) that separates the optimally controlled trajectory from the set (a cone) of suboptimal states at any given point. The interested readers are referred to [16, 17] and [18] for the original formulation directly from its authors; for a more modern treatment maintaining the same geometric intuition, we refer to [44] and the references therein.

To better illustrate this, suppose that $\widehat{\alpha}$ is the minimizer of the objective J in (1.3). Furthermore, making a slight abuse of notation, we denote by J^α the running cost of the objective $J(\alpha)$, i.e.

$$J_t^\alpha := \int_0^t f(X_s^\alpha, \alpha_s)\,\mathrm{d}s, \qquad 0 \le t \le T, \tag{1.4}$$

where X^α is a solution to (1.2) for the control α. From here it is clear that $J(\alpha) = J_T^\alpha + g(X_T^\alpha)$, with $J(\widehat{\alpha}) \le J(\alpha)$ for any arbitrary control α.

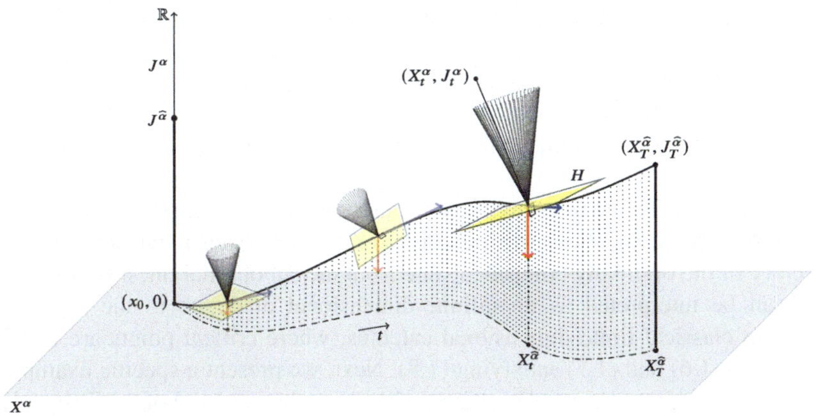

Fig. 1.1 Geometric interpretation of the PMP: the set of suboptimal actions form a family of cones in the (extended) phase space along the optimal trajectory $X^{\widehat{\alpha}}$ (the blue arrow pointing the direction of the state) pointing in the direction of decreasing cost (the red arrow pointing the direction of the costate), separated by a supporting hyperplane H (yellow plane)

Consider now the coordinate space (X^{α}, J^{α}) for every possible control α; see Fig. 1.1. In this extended phase space, it is possible to characterize $\widehat{\alpha}$ by observing that if a control α different from the optimal is chosen at the instant t, the corresponding (suboptimal) pair $(X_t^{\alpha}, J_t^{\alpha})$ not only necessarily breaks from the trajectory described by $(X^{\widehat{\alpha}}, J^{\widehat{\alpha}})$, but in fact the collection of suboptimal pairs forms a cone with apex $(X_t^{\widehat{\alpha}}, J_t^{\widehat{\alpha}})$, and therefore, there exists a separate hyperplane supporting the cone of suboptimal pairs through its apex:

$$\big(-(y, 1) \cdot \mathrm{d}(X_t^{\alpha}, J_t^{\alpha})\big) = -\big(y \cdot b(X_t^{\alpha}, \alpha_t)\big) - f(X_t^{\alpha}, \alpha_t) = \text{const.}, \tag{1.5}$$

where y is the corresponding costate variable in the X coordinate at time t.

From here, it is possible to recover the physical notion of the conservation of energy in Hamiltonian mechanics: Let H be defined as

$$H(x, y, a) := (y \cdot b(x, a)) + f(x, a),$$

and let Y be the linear conjugate of $X^{\alpha} - X^{\widehat{\alpha}}$ with the terminal condition $Y_T = \mathrm{D}_x\, g(X_T^{\widehat{\alpha}})$, where D_x denotes the Jacobian in the x axis. Since (1.5) describes a *supporting* hyperplane and $\widehat{\alpha}$ is optimal (i.e. $J(\widehat{\alpha}) - J(\alpha) \leq 0$), it follows from the classical calculus of variations arguments –recall the Euler-Lagrange equation characterizing minimizing smooth curves in the deterministic framework, see [37] or [27]– that $(X^{\widehat{\alpha}}, Y)$ solves the coupled system of differential equations.

$$\mathrm{d}\, X_t^{\widehat{\alpha}} = \mathrm{D}_y\, H(X_t^{\widehat{\alpha}}, Y_t, \widehat{\alpha}_t)\, \mathrm{d}t \quad (0 \leq t \leq T), \qquad X_0^{\widehat{\alpha}} = x_0, \tag{1.6}$$

$$\mathrm{d}\, Y_t = -\, \mathrm{D}_x\, H(X_t^{\widehat{\alpha}}, Y_t, \widehat{\alpha}_t)\, \mathrm{d}t \quad (0 \leq t \leq T), \qquad Y_T = \mathrm{D}_x\, g(X_T^{\widehat{\alpha}}), \tag{1.7}$$

where $\widehat{\alpha}$ is the control that satisfies the inequality

$$H(X_t^{\widehat{\alpha}}, Y_t, \widehat{\alpha}_t) \leq H(X_t^{\widehat{\alpha}}, Y_t, a), \qquad \forall a \in A, \tag{1.8}$$

for every $t \in [0, T]$. This inequality is known in the literature as the *Pontryagin's Condition*.

To summarize, the PMP establishes that an optimal control $\widehat{\alpha}$ is characterized by a set of differential equations (the conjugate or *adjoint* system) that "separates" the trajectory of the optimally controlled state from the suboptimal ones. Heuristically, PMP can be interpreted as an infinite-dimensional analogue to the "derivative test" from classical finite-dimensional calculus, where critical points are now the solutions to (1.6) and (1.7) satisfying (1.8). Next, we present a specific example of a deterministic optimal control problem to illustrate the potential of this Principle to solve this type of problems.

The version of the PMP for controlled stochastic processes was first studied by Bismut [14], and this fundamental extension represented interesting challenges. As can be seen in the system (1.6) and (1.7), it corresponds to a deterministic system of ordinary differential equations, where the first is a forward equation with a given initial condition, while the adjoint system is a backward equation for a given terminal condition. The existence of a solution adapted to the adjoint system and, in general, for backward stochastic differential equations has been an active area of research in recent decades; see, for example, the fundamental work of Pardoux and Peng [59] and Cadenillas and Karatzas [21].

1.2.1 An Illustrative Deterministic Example

The model, inspired by the one presented in [19] and the subsequent generalizations studied in [68], aims to maximize the utility of a global economy that produces a consumption good while simultaneously polluting the environment. For simplicity we shall consider only the finite horizon setting [0, 1], with a running linear payoff and a terminal utility $g \equiv 0$.

At any time $0 \leq t \leq 1$, we have the following nonnegative quantities:

K_t: Available capital;
P_t: Pollution index;
C_t: Consumption expenditure (portion of the capital that is spent);
θ_t: Abatement expenditure (portion of the capital that is invested on reducing pollution emissions).

Furthermore, we assume that the controller is subject to a prescribed limit of instantaneous expenditure:

$$(C_t, \theta_t) \in A := [0, \overline{C}] \times [0, \overline{\theta}], \qquad \forall\, t \in [0, 1],$$

for some fixed constants $0 < \overline{C} < \infty$ and $0 < \overline{\theta} < 1$. The controller aims to find a measurable strategy

$$(C, \theta) : [0, 1] \to A,$$

that maximizes the running payoff

$$\int_{[0,1]} U(C_t, P_t)\, dt, \qquad \text{with} \qquad U(c, p) := -\chi p + c,$$

where $\chi > 0$ represents the monetary cost per unit of pollution, subject to the controlled dynamics

$$K_t = K_0 + \int_{(0,t]} \{(1 - \theta_s)R_s - C_s\}\, ds,$$

$$P_t = P_0 + \int_{(0,t]} \{E_s - \pi P_s\}\, ds,$$

such that:

π : is the natural absorption rate of pollution by the environment, with $0 \le \pi < 1$,
R_t : represents the overall returns from the capital K_t, and
E_t : denotes the total emissions produced by a return R_t with an abatement investment $\theta_t R_t$.

Following [19], we assume a constant productivity factor $\gamma > 0$ for the return R_t, and that the emissions E_t increase proportionally to the overall returns R_t, and decrease proportionally to the abatement investment $\theta_t R_t$:

$$R_t = \gamma K_t \qquad\qquad E_t = \phi R_t - \sigma \theta_t R_t,$$

where $\phi > 0$ is the increment in the pollution index per unit of returns, and $\sigma > 0$ is the efficiency of abatement technology per unit of returns.

Let us now put all these pieces together, as the following optimal control problem:

$$\underset{(C,\theta)^\top:[0,1]\to A}{\text{maximize}} \int_{[0,1]} U(C_t, P_t)\, dt \tag{1.9}$$

where P_t is part of the solution to the ODE

$$\begin{pmatrix} K_t \\ P_t \end{pmatrix} = \begin{pmatrix} K_0 \\ P_0 \end{pmatrix} + \int_{(0,t]} \left\{ \gamma \begin{pmatrix} 1 - \theta_s & 0 \\ \phi - \sigma\theta_s & -\frac{\pi}{\gamma} \end{pmatrix} \begin{pmatrix} K_s \\ P_s \end{pmatrix} - \begin{pmatrix} C_s \\ 0 \end{pmatrix} \right\} ds. \tag{1.10}$$

To be consistent with our previous notation (cf. (1.2) and (1.3)), we shall henceforth write

$$X_t = \begin{pmatrix} K_t \\ P_t \end{pmatrix} \text{ and } \alpha_t = \begin{pmatrix} C_t \\ \theta_t \end{pmatrix},$$

as the state process and the control, respectively, rewriting succinctly (1.9) and (1.10) as

$$\underset{\alpha:[0,1]\to A}{\text{maximize}}\ J(\alpha), \qquad \text{where} \qquad J(\alpha) := \int_{[0,1]} f(X_t, \alpha_t)\,\mathrm{d}t, \tag{1.11}$$

such that

$$X_t = X_0 + \int_{(0,t]} b(X_s, \alpha_s)\,\mathrm{d}s, \tag{1.12}$$

with

$$\begin{aligned} b(x,a) :&= \gamma \begin{pmatrix} 1-\theta & 0 \\ \phi - \sigma\theta & -\frac{\pi}{\gamma} \end{pmatrix} x + \begin{pmatrix} 1\,0 \\ 0\,0 \end{pmatrix} a \\ &= \gamma \begin{pmatrix} 1 & 0 \\ \phi & -\frac{\pi}{\gamma} \end{pmatrix} x + \gamma k \begin{pmatrix} \frac{1}{\gamma k} & -1 \\ 0 & -\sigma \end{pmatrix} a \\ f(x,a) :&= \begin{pmatrix} 0 & -\chi \end{pmatrix} x + \begin{pmatrix} 1 & 0 \end{pmatrix} a \\ x &= \begin{pmatrix} k \\ p \end{pmatrix}, \ a = \begin{pmatrix} c \\ \theta \end{pmatrix}. \end{aligned} \tag{1.13}$$

In order to obtain the necessary conditions for optimality, suppose $\hat{\alpha}$ is a maximizer of (1.11), with $\hat{X} = X^{\hat{\alpha}}$ the corresponding controlled state. For any other control α and any subinterval $I^\epsilon \subset [0,1]$ of Lebesgue measure $\epsilon > 0$, we define the *spike variation* of $\hat{\alpha}$ –with respect to α and I^ϵ– as the (admissible) piecewise control

$$\alpha_t^\epsilon := \alpha_t \mathbf{1}_{I^\epsilon}(t) + \hat{\alpha}_t \mathbf{1}_{[0,1]\setminus I^\epsilon}(t),$$

with $X^\epsilon = X^{\alpha^\epsilon}$ the corresponding controlled state.

Remark 1.1 In the continuous deterministic (and convex A) case, it is possible to consider a “simpler” variation of the form

$$\alpha^\epsilon = \hat{\alpha} + \epsilon v, \text{ with } v \in \mathrm{C}^1([0,1]) \text{ and } v(0) = v(1) = 0,$$

which would allow us to use classical calculus of variations techniques. In our case, we will use the following version of Lyapunonv's convexity theorem to account for the lack of 'classical' differentiation, stating the condition, in its original version, under which a measurable set E of a measure different from zero must contain measurable subsets of arbitrary small measure.

Lemma 1.1 ([50], Lemma 1) *Let* $(\mathrm{X}, \mathcal{M}, \mu)$ *be a non-atomic finite measure space, and let*

$$\mathbf{w} = \Big\{ w = (w_i)_{1\le i\le n} \;\Big|\; \sum_{1\le i\le n} w_i \le 1,\ w_i \ge 0,\ \forall i = 1, \dots, n,\ \textit{and}\ n \in \mathbb{N} \Big\}.$$

For each $E \in \mathcal{M}$ *and* $w \in \mathbf{w}$*, there exists a family of measurable sets* $E_i(w)$*,* $1 \le i \le n$*, such that*

$$\mu\big(E_i(w)\big) = w_i \mu(E), \qquad \forall i = 1, \dots, n,$$

$$E_i(w) \cap E_j(w) = \emptyset \qquad \forall i, j = 1, \dots, n,\ \textit{with}\ i \neq j,$$

and

$$\mu\big(E_i(w) \triangle E_i(w^*)\big) \to 0 \qquad \textit{as} \qquad w \to w^*\ \textit{pointwise}.$$

In this problem, it means that for any $\epsilon > 0$ and any (scalar) Lebesgue integrable function l on $[0, 1]$ there exists a sub-interval $I^\epsilon \subset [0, 1]$ such that

$$\int_{I^\epsilon} l(s)\mathrm{d}s = \epsilon \int_{[0,1]} l(s)\mathrm{d}s + o(\epsilon). \tag{1.14}$$

Returning to the problem (1.11), w.l.o.g. let $I^\epsilon = (t_0, t_1]$ for some $t_0, t_1 \in [0, 1)$ with $t_1 - t_0 = \epsilon > 0$. From (1.12) we get the next estimate for the difference between the optimally controlled state $\hat{X}$ and X^ϵ:

$$\begin{aligned} |X_t^\epsilon - \hat{X}_t|^2 \le 2 &\left| \int_{(0,t]} \{b(X_s^\epsilon, \alpha_s^\epsilon) - b(\hat{X}_s, \hat{\alpha}_s)\} \mathbf{1}_{I^\epsilon} \mathrm{d}s \right|^2 \\ &+ 2 \left| \int_{(0,t]} \{b(X_s^\epsilon, \hat{\alpha}_s) - b(\hat{X}_s, \hat{\alpha}_s)\} \mathbf{1}_{[0,1]\setminus I^\epsilon} \mathrm{d}s \right|^2. \end{aligned} \tag{1.15}$$

For the first integral in the r.h.s. of (1.15), using Cauchy-Schwarz inequality, we have that

$$\left| \int_{(0,t]} \{b(X_s^\epsilon, \alpha_s^\epsilon) - b(\hat{X}_s, \hat{\alpha}_s)\} \mathbf{1}_{I^\epsilon} \mathrm{d}s \right|^2$$

$$\leq 2\epsilon^2 \int_{(0,t]} \left\{ \left| b(X_s^\epsilon, \alpha_s^\epsilon) - b(\hat{X}_s, \alpha_s^\epsilon) \right|^2 + \left| b(\hat{X}_s, \alpha_s^\epsilon) - b(\hat{X}_s, \hat{\alpha}_s) \right|^2 \right\} \mathrm{d}s.$$

On the one hand, from the linearity of b on X, applying Taylor's expansion for the first term yields the following estimation

$$\begin{aligned}
\left| b(X_s^\epsilon, \alpha_s^\epsilon) - b(\hat{X}_s, \alpha_s^\epsilon) \right|^2 &= \left| \int_{[0,1]} \mathrm{D}_x\, b(\hat{X}_s + \tau(X_s^\epsilon - \hat{X}_s), \alpha_s^\epsilon)(X_s^\epsilon - \hat{X}_s) \mathrm{d}\tau \right|^2 \\
&= \gamma^2 \left\| \begin{pmatrix} 1 - \theta_s & 0 \\ \phi - \sigma\theta_s & -\frac{\pi}{\gamma} \end{pmatrix} (X_s^\epsilon - \hat{X}_s) \right\|_2^2 \\
&\leq (\gamma^2 + \pi^2) \left| X_s^\epsilon - \hat{X}_s \right|^2.
\end{aligned}$$

On the other hand, for the second term we have the next estimation:

$$\begin{aligned}
\left| b(\hat{X}_s, \alpha_s^\epsilon) - b(\hat{X}_s, \hat{\alpha}_s) \right|^2 &= \left| \gamma \begin{pmatrix} \hat{\theta}_s - \theta_s & 0 \\ \sigma(\hat{\theta}_s - \theta_s) & 0 \end{pmatrix} \hat{X}_s + \begin{pmatrix} \hat{C}_s \\ 0 \end{pmatrix} - \begin{pmatrix} C_s \\ 0 \end{pmatrix} \right|^2 \\
&\leq 3\left\{ 2\gamma^2 \overline{\theta} |\hat{X}_s|^2 + |\hat{\alpha}_s|^2 + |\alpha_s|^2 \right\} \\
&\leq 3(2\gamma^2 + 1)\left\{ |\hat{X}_s|^2 + |\hat{\alpha}_s|^2 + |\alpha_s|^2 \right\}.
\end{aligned}$$

Then, applying Gronwall inequality, we conclude that

$$\left\| (X^\epsilon - \hat{X}) \mathbf{1}_{I^\epsilon} \right\|_\infty^2 \leq 6\epsilon^3 e^{2(\gamma^2 + \pi^2)} (2\gamma^2 + 1)\left\{ 1 + \|\hat{X}\|_\infty^2 + \|\hat{\alpha}\|_\infty^2 + \|\alpha\|_\infty^2 \right\}$$

for any α, with $\|f\|_\infty^2 = \sup_{t \in [0,1]} |f_t|^2$.

For the second term at the r.h.s of (1.15), since

$$(0, t] \cap ([0, 1] \setminus I^\epsilon) = (t_0, \min\{t, t_1\}] \cup (t_1, t],$$

with $(t_1, t] = \emptyset$ whenever $t < t_1$, observe that

- $X^\epsilon - \hat{X} = 0$ in $(t_0, \min\{t, t_1\}]$, and
- $X^\epsilon - \hat{X}$ solve the linear ODE

$$\frac{\mathrm{d}}{\mathrm{d}t}(X_t^\epsilon - \hat{X}_t) = -\gamma \begin{pmatrix} 1 & 0 \\ \phi & -\frac{\pi}{\gamma} \end{pmatrix} (X_t^\epsilon - \hat{X}_t), \qquad \forall t \in (t_1, t],$$

with initial condition $X_{t_1}^\epsilon - \hat{X}_{t_1}$ possibly distinct from 0; i.e.

$$(X_t^\epsilon - \hat{X}_t) \mathbf{1}_{[0,1] \setminus I^\epsilon} = \exp\left\{ \gamma \begin{pmatrix} 1 & 0 \\ \phi & -\frac{\pi}{\gamma} \end{pmatrix} (t - t_1) \right\} (X_{t_1}^\epsilon - \hat{X}_{t_1}), \quad \forall t \in (t_1, t].$$

Then, there exits a constant $C = C(\gamma, \phi, \pi, 1) > 0$ such that for every $t_1 \leq t \leq 1$,

$$\|(X^\epsilon - \hat{X})\mathbf{1}_{[0,1]\setminus I^\epsilon}\|_\infty^2 \leq C\|X_{t_1}^\epsilon - \hat{X}_{t_1}\|^2 \leq C\|(X^\epsilon - \hat{X})\mathbf{1}_{I^\epsilon}\|_\infty^2.$$

Consequently, from dominated convergence we have that the *first order variation* of $X^\epsilon - \hat{X}$ due to α^ϵ, defined as

$$\vartheta = \lim_{\epsilon \downarrow 0} \frac{X^\epsilon - \hat{X}}{\epsilon},$$

satisfies the linear ODE

$$\vartheta_t = \int_{(0,t]} \big\{ \mathrm{D}_x\, b(\hat{X}_s, \hat{\alpha}_s)\vartheta_s + \big(b(\hat{X}_s, \alpha_s) - b(\hat{X}_s, \hat{\alpha}_s)\big)\big\} \mathrm{d}s$$

by means of Lyapunov's convexity theorem; see (1.14). Moreover, we have

$$\|(X^\epsilon - \hat{X}) - \vartheta\|_\infty^2 = o(\epsilon^2).$$

Let us now move on to estimating the variation of the objective functional. For any control α we have the following.

$$J(\alpha) - J(\hat{\alpha}) = \int_{[0,1]} \big\{ f(X_t^\alpha, \alpha_t) - f(\hat{X}_t, \hat{\alpha}_t)\big\}\, \mathrm{d}t \leq 0. \tag{1.16}$$

However, from our previous estimates and the regularity of U, the variation in the cost functional due to a spike variation α^ϵ is of the form

$$f(X_t^\epsilon, \alpha_t^\epsilon) - f(\hat{X}_t, \hat{\alpha}_t) = f(\hat{X}_t + \varphi_t, \alpha_t^\epsilon) - f(\hat{X}_t, \hat{\alpha}_t) + o(\epsilon^2),$$

and, consequently,

$$\begin{aligned}
\frac{J(\alpha) - J(\hat{\alpha})}{\epsilon} &= \frac{1}{\epsilon} \int_{[0,1]} \big\{ f\big(\hat{X}_t + (X_t^\epsilon - \hat{X}_t), \alpha_t^\epsilon\big) - f(\hat{X}_t, \hat{\alpha}_t)\big\}\, \mathrm{d}t \\
&= \frac{1}{\epsilon} \int_{[0,1]} \{ f(\hat{X}_t + \varphi_t, \alpha_t^\epsilon) - f(\hat{X}_t, \alpha_t^\epsilon)\} \mathrm{d}t \\
&\quad + \frac{1}{\epsilon} \int_{[0,1]} \big\{ f(\hat{X}_t, \alpha_t^\epsilon) - f(\hat{X}_t, \hat{\alpha}_t)\big\} \mathbf{1}_{I^\epsilon}\, \mathrm{d}t + (\text{lower order terms}).
\end{aligned}$$

This means that the first order variation of the objective functional is

$$\begin{aligned}\frac{\mathrm{d}}{\mathrm{d}\alpha}J(\alpha) &= \lim_{\epsilon\downarrow 0}\frac{J(\alpha)-J(\hat{\alpha})}{\epsilon}\\ &= \int_{[0,1]}\left\{\mathrm{D}_x f(\hat{X}_t,\hat{\alpha}_t)\vartheta_t + \left(f(\hat{X}_t,\alpha_t)-f(\hat{X}_t,\hat{\alpha}_t)\right)\right\}\mathrm{d}t.\end{aligned}\tag{1.17}$$

Now, we introduce the adjoint process Y as the orthogonal of ϑ in L^2:

$$\int_{[0,1]}\left\{Y_t^\top\left(\frac{\mathrm{d}}{\mathrm{d}t}\vartheta_t - \mathrm{D}_x b(\hat{X}_t,\hat{\alpha}_t)\vartheta_t - \left(b(\hat{X}_t,\alpha_t)-b(\hat{X}_t,\hat{\alpha}_t)\right)\right)\right\}\mathrm{d}t = 0.\tag{1.18}$$

Integrating by parts, we obtain

$$\int_{[0,1]}\left\{Y_t^\top\left(\frac{\mathrm{d}}{\mathrm{d}t}\vartheta_t\right)\right\}\mathrm{d}t = \left(Y_T^\top\vartheta_T - \underbrace{Y_0^\top\vartheta_0}_{=0}\right) - \int_{[0,1]}\left\{\left(\frac{\mathrm{d}}{\mathrm{d}t}Y_t\right)^\top\vartheta_t\right\}\mathrm{d}t;\tag{1.19}$$

thus, by subtracting (1.18) from (1.17) and substituting with (1.19) where necessary, we get the following:

$$\begin{aligned}\frac{\mathrm{d}}{\mathrm{d}\alpha}J(\alpha) &= -Y_T^\top\vartheta_T + \int_{[0,1]}\left\{\left(\frac{\mathrm{d}}{\mathrm{d}t}Y_t\right)^\top + \mathrm{D}_x f(\hat{X}_t,\hat{\alpha}_t) + Y_t^\top\,\mathrm{D}_x b(\hat{X}_t,\hat{\alpha}_t)\right\}\vartheta_t\mathrm{d}t\\ &\quad + \int_{[0,1]}\left\{\left(f(\hat{X}_t,\alpha_t)-f(\hat{X}_t,\hat{\alpha}_t)\right) + Y_t^\top\left(b(\hat{X}_s,\alpha_s)-b(\hat{X}_s,\hat{\alpha}_s)\right)\right\}\mathrm{d}t\\ &\le 0,\end{aligned}$$

where the last inequality is due to the optimality of $\hat{\alpha}$; see (1.16).

We now define the function

$$H(x,y,a) := f(x,a) + y^\top b(x,a)$$

and refer to it as the *Hamiltonian* of the problem. Then, plugging H into our previous computation yields the variational inequality

$$\begin{aligned}\frac{\mathrm{d}}{\mathrm{d}\alpha}J(\alpha) = -Y_1^\top\vartheta_1 + &\int_{[0,1]}\left\{\left(\frac{\mathrm{d}}{\mathrm{d}t}Y_t\right)^\top + \mathrm{D}_x H(\hat{X}_t,Y_t,\hat{\alpha}_t)\right\}\vartheta_t\mathrm{d}t\\ &+ \int_{[0,1]}\left\{H(\hat{X}_t,Y_t,\alpha_t) - H(\hat{X}_t,Y_t,\hat{\alpha}_t)\right\}\mathrm{d}t \le 0\end{aligned}$$

for any arbitrary control α. This means that for $\hat{\alpha}$ to be optimal, it must necessarily hold that:

- X solves (1.6); equivalently, K and P solve the system of ODEs with initial conditions

$$\left.\begin{aligned} \mathrm{d}K_t &= \big\{(1-\hat{\theta}_t)\gamma K_t - \hat{C}_t\big\}\,\mathrm{d}t, \\ \mathrm{d}P_t &= \big\{(\phi-\sigma\hat{\theta}_t)\gamma K_t - \pi P_t\big\}\,\mathrm{d}t, \end{aligned}\right\} \qquad \forall t \in (0,1], \tag{1.20}$$

  with $K_0 = k > 0$ and $P_0 = p > 0$.
- Y solves (1.7), ensuring that the equality $\frac{\mathrm{d}}{\mathrm{d}\alpha} J(\alpha) = 0$ holds at the optimal $\hat{\alpha}$. In terms of the parameters of the model, we have that $Y^{(1)}$ and $Y^{(2)}$ solve the system of ODEs with terminal conditions

$$\left.\begin{aligned} \mathrm{d}Y_t^{(1)} &= -\gamma\big\{(1-\hat{\theta}_t)Y_t^{(1)} + (\phi-\sigma\hat{\theta}_t)Y_t^{(2)}\big\}\,\mathrm{d}t, \\ \mathrm{d}Y_t^{(2)} &= \big\{\chi + \pi Y_t^{(2)}\big\}\,\mathrm{d}t, \end{aligned}\right\} \qquad \forall t \in [0,1), \tag{1.21}$$

  with $Y_1^{(1)} = Y_1^{(2)} = 0$.
- H satisfies for each $t \in [0,1]$ the variational inequality

$$H(\hat{X}_t, Y_t, a) \leq H(\hat{X}_t, Y_t, \hat{\alpha}_t) \qquad \forall a \in A, \tag{1.22}$$

  also known as the *maximum principle* or *Pontryagin's condition*.

The basic reasoning behind the derivation of these conditions is that, given a candidate optimal control $\hat{\alpha}$, the first-order variation of the performance measure should be non-positive for 'small' variations of $\hat{\alpha}$.

As indicated above, the previous results provide us with the necessary conditions for control optimization. However, the results can be extended to cover sufficient conditions for optimality. Indeed, in addition to conditions (1.20)–(1.22), for every $t \in [0,1]$ define

$$x \longmapsto \max_{a\in A} H(x, Y_t, a) =: \mathcal{H}_t(x).$$

From the linearity of the coefficients (1.13) we have that $\mathcal{H}$ is linear on x, and thus concave. Moreover, from (1.22) we also have that the candidate for optimal control $\hat{\alpha}$ lies in the corners of A, and is determined by the gradient of the Hamiltonian. Notice that

$$\mathrm{D}_a\, H(\hat{X}_t, Y_t, \hat{\alpha}_t) = \begin{pmatrix}1 & 0\end{pmatrix} + Y_t^{\top}\begin{pmatrix}1 & -\gamma\hat{X}_t^{(1)} \\ 0 & -\sigma\gamma\hat{X}_t^{(1)}\end{pmatrix};$$

then, the optimal control candidate is given by

$$\hat{\alpha}_t^{(1)} = \hat{C}_t := \begin{cases} 0, & \text{if } Y_t^{(1)} \leq -1, \\ \overline{C}, & \text{if } Y_t^{(1)} > -1, \end{cases} \quad \text{and } \hat{\alpha}_t^{(2)} = \hat{\theta}_t := \begin{cases} 0, & \text{if } -\sigma Y_t^{(2)} \leq Y_t^{(1)}, \\ \overline{\theta}, & \text{if } -\sigma Y_t^{(2)} > Y_t^{(1)}. \end{cases} \tag{1.23}$$

Optimal controls operate in a pure bang-bang mode, alternating between maximum consumption and maximum reduction based on the evolution of the trajectory Y.

We now proceed to prove the optimality of $\hat{\alpha}$. On the one hand, observe that using integration by parts, we obtain

$$\begin{aligned} 0 &= Y_T^\top (X_T^\alpha - \hat{X}_T) - Y_0^\top (X_0^\alpha - \hat{X}_0) \\ &= \int_{[0,1]} \{Y_t^\top \big(b(X_t^\alpha, \alpha_t) - b(\hat{X}_t, \hat{\alpha}_t)\big) - \mathrm{D}_x H(\hat{X}_t, Y_t, \hat{\alpha}_t)^\top (X_t^\alpha - \hat{X}_t)\}\mathrm{d}t. \end{aligned}$$

On the other hand, from the definition of $\mathcal{H}$, we get that, for any control α,

$$H(x, Y_t, \alpha_t) \leq \mathcal{H}(x, Y_t), \qquad \forall x \in \mathbb{R}^2,$$

with an equality at $(X^{\hat{\alpha}}, \hat{\alpha})$ due to the maximum principle. Thus, from the concavity of $\mathcal{H}$,

$$\begin{aligned} J(\alpha) - J(\hat{\alpha}) &= \int_{[0,1]} \{f(X_t^\alpha, \alpha_t) - f(\hat{X}_t, \hat{\alpha}_t)\}\,\mathrm{d}t + 0 \\ &= \int_{[0,1]} \{\big(H(X_t^\alpha, Y_t, \alpha_t) - H(\hat{X}_t, Y_t, \hat{\alpha}_t)\big) - \mathrm{D}_x H(\hat{X}_t, Y_t, \hat{\alpha}_t)^\top (X_t^\alpha - \hat{X}_t)\}\,\mathrm{d}t \\ &\leq \int_{[0,1]} \{\big(H(X_t^\alpha, Y_t, \alpha_t) - H(\hat{X}_t, Y_t, \hat{\alpha}_t)\big) - \big(\mathcal{H}(X_t^\alpha, Y_t) - \mathcal{H}(\hat{X}_t, Y_t)\big)\}\,\mathrm{d}t \\ &\leq 0, \end{aligned}$$

proving the optimality of the control $\hat{\alpha}$, defined in (1.23).

1.3 Stochastic Analysis on Point Processes

We present the basic concepts and results needed for the applied calculus of jump diffusions. We will just briefly review the basic theory here, as there are several excellent books which provide a complete account of the details. The following section is based primarily on [29, 45, 46] and [60], which are recommended for those readers interested in further study. In particular, we will follow the general treatment and naming conventions presented by Kingman [45, 46].

Let $(\Omega, \mathcal{F}, \mathbb{P})$ be a given probability space, and let $(S, \mathcal{B}(S))$ be a measurable space, which we will call the *state space*. For pedagogical purposes we will assume that $S \subset \mathbb{R}^d$ is measurable, with $\mathcal{B}(S)$ its Borel σ-algebra; alternatively, we need to assume the diagonal

$$\{(x, y) \in S \times S \,|\, x = y\}$$

is measurable in $(S \times S, \mathcal{B}(S) \otimes \mathcal{B}(S))$. Then, a *point process* Π on S is simply defined as a random countable subset of S:

$$\omega \longmapsto \Pi(\omega) \subseteq S \text{ measurable, with } \#\Pi(\omega) \text{ countable.}$$

Observe that a point process Π naturally induces a collection of random variables

$$N(B) = \#(\Pi \cap B), \qquad\qquad \text{for } B \in \mathcal{B}(S),$$

which *count* the number of elements, or points, present at some given section of Π. Moreover, since $\Pi(\omega)$ is countable by definition, the family $N = \{N(B), B \subseteq S\}$ as a whole can be regarded as an *integer-valued random measure*.

To be precise, N is a measurable mapping

$$N : (\Omega, \mathcal{F}) \longrightarrow (\mathcal{M}_c^*, \mathcal{B}(\mathcal{M}_c^*)),$$

where $\mathcal{M}_c^* \equiv \mathcal{M}_c^*(S)$ denotes the space of $\mathbb{Z}_+ \cup \{\infty\}-$valued, $\sigma-$finite measures defined over $\mathcal{B}(S)$, endowed with the Borel $\sigma-$algebra $\mathcal{B}(\mathcal{M}_c^*)$. The elements of $\mathcal{M}_c^*$ will be referred to as *counting measures*.

Of particular importance are the so-called *Poisson point processes* due to their flexibility and usefulness in both theoretical and practical applications.

Definition 1.1 ([46]) A *Poisson point process* Π on S is a point process that satisfies the following conditions:

1. for any collection of disjoint measurable sets $\{B_1, \ldots, B_n\}$ of S, the random variables $N(B_1), \ldots N(B_n)$ are independent;
2. for any $B \in \mathcal{B}(S)$, the random variable $N(B)$ is distributed Poisson of parameter $\Lambda(B) \in \mathbb{R}_+ \cup \{\infty\}$, with the particular case of

$$\mathbb{P}[N(B) = \infty] = 1 \quad \text{if} \quad \Lambda(B) = \infty.$$

Consequently, any counting measure N satisfying the previous two conditions is known as a *Poisson random measure*.

Remark 1.2 The appearance of the Poisson distribution in the previous definition is not arbitrary. In words of Kingman, "it is an inevitable consequence of the 'complete randomness' inherent in our assumptions of independence"; to be more precise, one can see that under the stated independence conditions, the expected infinitesimal increments of test sets satisfy a system of Poisson ordinary differential equations. Heuristically, one need only to recall the role the Poisson law plays in the well-known *law of rare events* from elementary probability (see [67]), or even the Gaussian law in the *central limit theorem*. These arguments, while technically simple, do not contribute much to our current discussion; hence, we encourage any interested reader to check Section 1.4 in [45], "The inevitability of the Poisson distribution".

Observe that it is technically possible to introduce integer-valued random measures without the need to have an underlying point process; see, for instance, [28]. The key but subtle difference lies in the fact that point processes emphasize the *configuration of points (their locations)*, while integer-valued measures emphasize their *counts (integration against test functions)*. In this sense, the measure-focused perspective can be considered broader; however, the points-focused perspective on itself produces a rich theory on itself, which allows us to work with many interesting examples, as those presented in the next subsection.

From Definition 1.1, notice that the mapping

$$B \longmapsto \Lambda(B) \equiv \mathbb{E}[N(B)], \qquad \forall B \in \mathcal{B}(S), \tag{1.24}$$

defines a measure over $(S, \mathcal{B}(S))$; we henceforth refer to Λ as the *intensity* or *mean measure* of N. Furthermore, we refer to the signed random measure

$$\tilde{N}(B) := N(B) - \Lambda(B), \qquad \forall B \in \mathcal{B}(S),$$

as the *compensated counting measure of* N. In the literature there are several ways to approach the role of the intensity measure within the theory of stochastic processes, but for the sake of brevity, we will present only what is necessary to follow this text.

First, it must be highlighted that a key property from intensity measures is that they can guarantee the existence of point processes, provided a weak finiteness condition holds (particularly, σ−finiteness); see Theorem 2 in [46]. Second, they provide ways to construct new point processes from previous ones, as we describe next: Suppose that Π_n, $n \geq 1$, is a sequence of independent Poisson point processes on S, each with an intensity measure Λ_n; then, the union

$$\bigcup_{n\geq 1} \Pi_n \text{ is a Poisson point process with intensity measure } \sum_{n\geq 1} \Lambda_n.$$

This result is known in literature as the *superposition property*. Conversely, if Π, is a Poisson point processes on S with intensity measure Λ, and S_1 is a measurable subset of S, then

$$\Pi \cap S_1 \text{ is a Poisson point process with intensity measure } \Lambda(\,\cdot\, \cap S_1).$$

This is known in literature as the *thinning property*.

For a more involved construction, let Π be a Poisson point process on S with intensity measure Λ. Let $(\mathbf{R}, \mathcal{R})$ be another measurable (Polish) space, and consider a *transition kernel* K from $(S, \mathcal{B}(S))$ into $(\mathbf{R}, \mathcal{R})$; that is,

$$K : S \times \mathcal{R} \to \mathbb{R} \tag{1.25}$$

such that

- $K(x, \cdot)$ is a probability measure on $(\mathbf{R}, \mathcal{R})$ for each $x \in S$, and
- $K(\cdot, C)$ is a measurable function for each $C \in \mathcal{R}$.

Then, a *marking of* Π –denoted by Π^*– is simply a random subset in the product space $S \times \mathbf{R}$,

$$\omega \mapsto \Pi^*(\omega) = \{(X(\omega), r_{X(\omega)}) \mid X(\omega) \in \Pi(\omega),\ r_{X(\omega)} \in \mathbf{R}\} \subset S \times \mathbf{R}$$

whose projection onto S is Π, and such that the conditional distribution of Π^* given Π makes r_X (or *marks*) independent, with the distribution given by $K(X, \cdot)$.

Theorem 1.1 (Marking Theorem, Sec. 5.2 in [45]) *The random subset Π^* is a Poisson point process on $S \times \mathbf{R}$ with intensity measure*

$$\Lambda^*(C) = \int_C \Lambda(\mathrm{d}x) \otimes K(x, \mathrm{d}r), \quad C \in \mathcal{B}(S) \times \mathcal{R}.$$

Remark 1.3 This decomposition of the expected (product) measure in terms of the stochastic kernel K and the (expected) measure Λ, similar to the decomposition of product measures, will be one of the main elements in the jump models presented in the next chapters, where $r \equiv r_X$ represent the marks that influence the evolution of the rest of the state; see (2.5).

Focusing for a moment on the induced counting measures, from the proper representation of Π, we have that for any measurable function $f : S \to \mathbb{R}$, the integral of f with respect to N is the random variable

$$\omega \mapsto \sum_{X \in \Pi(\omega)} f(X) =: \Big(\int f(x) N(\mathrm{d}x)\Big)(\omega).$$

The well-posedness of this integral is captured in the following celebrated theorem due to Campbell, and which proof can be found [45]:

Theorem 1.2 (Campbell's Theorem) *Let Π be a Poisson point process on S with intensity measure Λ, and let $f : S \to \mathbb{R}$ be measurable. Then the sum*

$$\mathbf{\Sigma}_f := \sum_{X \in \Pi} f(X)$$

is absolutely convergent with probability one if and only if

$$\int_S \min\{|f(x)|, 1\}\,\Lambda(\mathrm{d}x) < \infty. \tag{1.26}$$

If this condition holds, then

$$\mathbb{E}\big[\exp\{\zeta\,\boldsymbol{\Sigma}_f\}\big] = \exp\left\{\int_S \{e^{\zeta f(x)} - 1\}\Lambda(\mathrm{d}x)\right\} \tag{1.27}$$

for any $\zeta \in \mathbb{C}$ for which the integral on the right is finite, and, in particular, whenever ζ has no real part. Moreover,

$$\mathbb{E}[\boldsymbol{\Sigma}_f] = \int_S f(x)\Lambda(\mathrm{d}x) \tag{1.28}$$

in the sense that the expectation exists if and only if the integral exists (finite or infinite), and they are equal. If (1.28) *is finite, then*

$$\mathrm{Var}[\boldsymbol{\Sigma}_f] = \int_S f(x)^2\Lambda(\mathrm{d}x),$$

exists, and has the same value as the integral in the r.h.s. (finite or infinite).

1.3.1 Marked Poisson Processes Over $S = (0, \infty)$

In order to illustrate some of the evolutionary properties of point processes, consider for a moment the marked Poisson point process Π over $(0,\infty)\times\mathbb{R}$ of intensity measure $\Lambda(\mathrm{d}t, \mathrm{d}x) = \lambda(t)\,\mathrm{d}t \otimes \nu(\mathrm{d}x)$, for some integrable function $\lambda \geq 0$ and a probability measure ν. Then, by taking the test sets $\{(0,t]\times B \mid t \geq 0, B \in \mathcal{B}(\mathbb{R})\}$ we can see that for each B, the stochastic process is defined, for $t \geq 0$, as

$$N_t^B := N((0,t]\times B),$$

is a process with independent increments (since the intervals $(0,t]$ and $(t, t+h]$ are disjoint, their respective counting measures are by definition independent), null at zero (under the convention of $(0,0] = \emptyset$), and its distribution at time t is Poisson with parameter

$$\mathbb{E}[N_t^B] = \int_{(0,t]} \nu(B)\lambda(s)\,\mathrm{d}s.$$

That is, each N^B is a non-homogeneous Poisson process of rate $\nu(B)\lambda(t)$. Moreover, by setting

$$Y_t := \int_{(0,t]\times\mathbb{R}} x\, N(\mathrm{d}t, \mathrm{d}x), \qquad \forall t \geq 0, \tag{1.29}$$

and writing the explicit points of Π as $\{(\tau_n, \xi_n)\}_{n\geq 1}$, we have that

$$Y_t = \sum_{n\geq 1} \xi_n \mathbf{1}_{\{0<\tau_n\leq t\}} = \sum_{n=1}^{N_t^{\mathbb{R}}} \xi_n;$$

i.e. $\{Y_t\}$ is a compound Poisson process with jumps of size $\xi \sim \nu$, driven by an non-homogeneous Poisson process $N^{\mathbb{R}}$ of rate $\nu(\mathbb{R})\lambda(t) = 1 \cdot \lambda(t)$. This particular type of processes will be of interest in the following sections.

Let us now consider the case when $\mathbf{R}$ is a Polish space, and let Π be a Poisson point process on $S = (0, \infty) \times \mathbf{R}$ with intensity measure given by

$$\Lambda(\mathrm{d}t, \mathrm{d}r) = \lambda(t, r)\,\mathrm{d}t \otimes \nu(\mathrm{d}r), \tag{1.30}$$

with $\lambda : [0, \infty) \times \mathbf{R} \to [0, \infty)$ a measurable and bounded function, and ν a σ–finite measure on $(\mathbf{R}, \mathcal{R})$. Then, it can be verified –see Proposition 9.3.X in [29]– that

$$\mathbb{P}[N(\{t\} \times \mathbf{R}) \in \{0, 1\}] = 1 \qquad \forall t \geq 0 \tag{1.31}$$

Those marked Poisson processes satisfying (1.31) are referred to as *simple*.

The above result describes not only the behavior of the points that appear, but also the variation of the process as a whole. Indeed, recall that the *quadratic variation* of a stochastic process $X = \{X_t\}_{t\geq 0}$ is defined as

$$[X]_t = \lim_{\|P\|\to 0} \sum_{t_k\in P} (X_{t_k} - X_{t_{k-1}})^2,$$

where P is any finite partition of the interval $[0, t]$.

Lemma 1.2 *Let Π be a simple point process. Then,*

$$[N_\cdot^B]_t = N_t^B, \tag{1.32}$$

where

$$N_t^B := N((0, t] \times B), \qquad \forall t \geq 0, \forall B \in \mathcal{R}. \tag{1.33}$$

Proof Let N^B be as above, for an arbitrary $B \in \mathcal{R}$. From the definition of quadratic variation,

$$
\begin{aligned}
[N_\cdot^B]_t &= \lim_{\|P\|\to 0} \sum_{t_k \in P} \big(N((0,t_k]\times B) - N((0,t_{k-1}]\times B)\big)^2 \\
&= \lim_{\|P\|\to 0} \sum_{t_k \in P} \big(N((t_{k-1},t_k]\times B)\big)^2 \\
&= \lim_{\|P\|\to 0} \sum_{t_k \in P} N((t_{k-1},t_k]\times B) = N((0,t]\times B),
\end{aligned}
$$

where the second to last equality is due to (1.31). □

In what follows, we recall some well-known results from the general theory of stochastic processes, in order to define properly the stochastic integral and the compensator. Let $(\Omega, \mathcal{F}, \mathbb{F}, \mathbb{P})$ a *filtered probability space*, with filtration $\mathbb{F} = \{\mathcal{F}_t\}_{t\geq 0}$. A filtration is said to be *right-continuous* (resp. left-continuous) if $\mathcal{F}_t = \mathcal{F}_{t+}$ (resp. $\mathcal{F}_t = \mathcal{F}_{t-}$), for all $t \geq 0$, where

$$
\mathcal{F}_{t+} := \bigcap_{s>t} \mathcal{F}_s \quad \text{and} \quad \mathcal{F}_{t-} := \sigma\left(\bigcup_{s<t} \mathcal{F}_s\right).
$$

In fact, the filtration $\{\mathcal{F}_{t+}\}_{t\geq 0}$ is itself right-continuous. In the particular case of the Poisson point process Π in $(0,\infty)\times S$ with an intensity measure $\Lambda(\mathrm{d}t \times \mathrm{d}x)$, the *canonical filtration associated with* N is defined as

$$
\mathcal{F}_t^N := \sigma\{N(B\times C) \mid B \in \mathcal{B}((0,t]), C \in \mathcal{B}(S)\},
$$

A function $\tau : \Omega \to [0,\infty]$ is called a *stopping time* with respect to the filtration $\mathbb{F}$ if

i. τ is $\mathcal{F}_\infty$-measurable, where $\mathcal{F}_\infty = \sigma\,(\mathcal{F}_t, t \geq 0)$;
ii. $\{\tau \leq t\} \in \mathcal{F}_t$ for all $t \geq 0$.

Given an $\mathbb{F}$–adapted stochastic process $\{X_t\}$, the *hitting time* (or first entrance time) of a set A is defined as

$$
T_A(\omega) = \begin{cases} \inf\{t \geq 0 : X_t(\omega) \in A\}, & \text{if } \{t : X_t(\omega) \in A\} \neq \emptyset; \\ \infty, & \text{if } \{t : X_t(\omega) \in A\} = \emptyset. \end{cases}
$$

In fact, when X has right-continuous paths and A is open, T_A is a stopping time. If X has continuous paths and A is closed, then T_A is a stopping time.

The notion of hitting time is specially useful for the processes we are working with, as they allow us to determine the distribution of both the jumping times and the marks themselves. This is the idea behind the following result.

Lemma 1.3 *Let* $\Pi = \{(t, \Delta_t)\}$ *be a Poisson point process over* $(0,\infty)\times \mathbf{R}$, *defined on a probability space* $(\Omega, \mathcal{F}, \mathbb{P})$, *with intensity measure* $\Lambda(\mathrm{d}t, \mathrm{d}r) = \mathrm{d}t \otimes \nu(\mathrm{d}r)$.

Let $B \in \mathcal{R}$ with $0 < \nu(B) < \infty$, and hitting time

$$T_B := \inf\{t \geq 0 : \Delta_t \in B\}.$$

Then, T_B is independent of Δ_{T_B}. Moreover, T_B is exponentially distributed with parameter $\nu(B)$, and the law of Δ_{T_B} is given by

$$\frac{\nu(\cdot \cap B)}{\nu(B)}.$$

Proof Given $A \in \mathcal{R}$, with $A \subset B$, it is clear that

$$\mathbb{P}(T_B \leq t,\ \Delta_{T_B} \in A) = \mathbb{P}(T_A \leq T_{B\setminus A},\ T_A \wedge T_{B\setminus A} \leq t).$$

On the other hand, the random variable T_A is the first jump time of the Poisson process

$$N_t^A = \text{card}\{s \leq t : \Delta_s \in A\}.$$

Thus, $T_A \sim \text{Exp}(\nu(A))$. Since A and $B\setminus A$ are disjoint, T_A and $T_{B\setminus A}$ are independent random variables, and $T_{B\setminus A} \sim \text{Exp}(\nu(B) - \nu(A))$. Therefore, integrating the density, it follows that

$$\mathbb{P}(T_B \leq t,\ \Delta_{T_B} \in A) = \frac{\nu(A)}{\nu(B)}\big(1 - e^{-\nu(B)t}\big).$$

□

Another important feature that relates the point processes with the theory of martingales and stochastic integration is the *compensator*.

Definition 1.2 Let Π be a Poisson point process on $(0, \infty) \times \mathbf{R}$ with counting measure N. An $\mathbb{F}$–compensator for N is defined as the (random) measure A on $(0, \infty) \times \mathbf{R}$ such that for every $B \in \mathcal{R}$,

- the map $t \mapsto A((0, t] \times B)$ is a monotonic nondecreasing right-continuous $\mathbb{F}$–predictable process, and
- for each $n \geq 1$, the stopped process

$$t \longmapsto N((0, T_n \wedge t] \times B) - A((0, T_n \wedge t] \times B)$$

is an $\mathbb{F}$–martingale, where T_n is the hitting time

$$T_n(\omega) = \inf\{t \geq 0\ :\ N((0, t] \times B) \leq n\}.$$

Next result, presented without a proof, establishes a one-to-one correspondence between simple point processes and their compensator; see Theorem 14.1.IV of [29] and its corollaries.

Proposition 1.1 *Let* Π *be a simple point process such that the conditional distribution of interarrival times* $G_n(x) = \mathbb{P}[T_{n+1} - T_n \leq x|\mathcal{F}_{T_n}]$ *has a regular version and* $1 - G_n(x) > 0$ *for all* $x > 0$. *Then, the point process defines the compensator uniquely, and, conversely, the compensator uniquely defines the process.*

Then, from (Campbell's) Theorem 1.2, we have the following identification. Let Π be a Poisson point process on $(0, \infty) \times \mathbf{R}$ with intensity Λ as in (1.30). Then, for any arbitrary $B \in \mathcal{R}$, the following processes are $\mathbb{F}^N$–martingales:

1. $t \longmapsto N((0, t] \times B) - \Lambda((0, t] \times B) := \tilde{N}((0, t] \times B)$.
2. $t \longmapsto \tilde{N}((0, t] \times B)^2 - \Lambda((0, t] \times B)$.
3. $t \longmapsto \exp\{iuN((0, t] \times B) - \int (e^u - 1)\mathbf{1}_{(0,t]\times B}\, \Lambda(\mathrm{d}t, \mathrm{d}r)\}$, for any $u \in \mathbb{R}$.

From this one can see that two Π_1 and Π_2 Poisson point processes are independent if and only if their counting measures N_1 and N_2 do not jump simultaneously. For our purposes, we are interested in the following alternative characterization of the compensator, equivalent to the martingale characterization, but in terms of stochastic integration; see [29], Theorem 14.2.II.

Proposition 1.2 *Let* Π *be a Poisson point process on* $(0, \infty) \times \mathbf{R}$ *with counting measure* N. *Its* $\mathbb{F}$*–compensator* A *is defined as the unique* $\mathbb{F}$*–predictable measure* A *such that*

$$\mathbb{E}\Big[\int_{(0,\infty)\times\mathbf{R}} H(t, r)N(\mathrm{d}t, \mathrm{d}r)\Big] = \mathbb{E}\Big[\int_{(0,\infty)\times\mathbf{R}} H(t, r)A(\mathrm{d}t, \mathrm{d}r)\Big]$$

for every nonnegative $\mathbb{F}$*–predictable process* H.

It is on account of this relation that the compensator is referred to as the dual predictable projection in literature that adheres to the terminology of the general theory of processes. The use of the word 'projection' relates to the fact that any conditional expectation can be viewed as a projection from one space of functions to a subspace of functions having a coarser structure, see the general discussion of Chapter 14 at [29]. Lastly, we highlight the fact that encoded in the definition of Poisson point process/random measure there is the implicit notion *independent increments*, which will play an important role in the next section.

Chapter 2
FBSDEs with Environment-Dependent Jumps

Abstract The aim of this chapter is to present a coupled system of forward-backward stochastic differential equations, analyzing the existence and uniqueness of a strong solution. The main role throughout is played by a random environment process taking values in the Wasserstein space, influencing the jump-rate function of the backward and forward components of the system. The argument is based on the Banach fixed-point theorem, once the proper spaces are built, together with the operator that is iterated.

2.1 Introducing Random Environments

In this chapter, we consider *forward-backward stochastic differential equations* (FBSDEs in short) of the form

$$
\begin{aligned}
X_t^\mu =& X_0^\mu + \int_0^t b_s^\mu\left(X^\mu, Y^\mu, Z^\mu, U^\mu\right) \mathrm{d}t + \int_0^t \sigma_s^\mu\left(X^\mu, Y^\mu, Z^\mu, U^\mu\right) \mathrm{d}W_t \\
&+ \int_{(0,t]\times\mathbf{R}} \gamma_s^\mu\left(r, X^\mu, Y^\mu, Z^\mu, U^\mu\right) \tilde{\eta}^\lambda(\mathrm{d}s, \mathrm{d}r),
\end{aligned} \tag{2.1}
$$

$$
\begin{aligned}
Y_t^\mu =& Y_T^\mu + \int_t^T f_s^\mu\left(X^\mu, Y^\mu, Z^\mu, U^\mu\right) \mathrm{d}s - \int_t^T Z_s^\mu \,\mathrm{d}W_s \\
&- \int_{(t,T]\times\mathbf{R}} U_s^\mu(r)\tilde{\eta}^\lambda(\mathrm{d}s, \mathrm{d}r) - (M_T^\mu - M_t^\mu),
\end{aligned} \tag{2.2}
$$

$$
X_0^\mu \sim \mu_0 \in \mathcal{P}_2(\mathbb{R}^d; \mathcal{F}_0), \qquad Y_T^\mu = g_T^\mu(X^\mu) \in L^2(\mathbb{R}^n; \mathcal{F}_T), \qquad M_0 = 0. \tag{2.3}
$$

Throughout, we let $(\Omega, \mathcal{F}, \mathbb{F}, \mathbb{P})$ be a particular probability filtered space, to be specified later, on which a Brownian motion W, a compensated random measure associated with a marked Poisson process $\tilde{\eta}^\lambda$, and a random environment μ are

D. Hernández-Hernández, J. H. Ricalde-Guerrero, *Optimal Control in Random Environments*, SpringerBriefs in Mathematics,
https://doi.org/10.1007/978-3-032-25641-6_2

defined. We aim to establish the existence and uniqueness of (strong) solutions (X, Y, Z, U, M) to (2.1)–(2.3), where M is a compensating martingale term. The probability space is important in order to have a well-defined solution, and then we borrow the construction of the Wiener-Poisson space, guaranteeing that some version of the martingale representation theorem holds; see Section III.4 in [43] and Definition 2.2 below.

One of the main features of the model consists of an auxiliary, square-integrable process μ taking values in the Wasserstein space $(\mathcal{P}_2(\mathbb{R}^d), \mathcal{W}_2)$. This process plays the role of a random environment influencing the dynamics of the system. For the time being we will only consider μ as a possibly *exogenous* random environment in order to establish sufficient conditions for the existence and uniqueness of solutions to fully-coupled FBSDEs. A particular case of an *endogenous* random environment will be treated in Chap. 4.

Specific assumptions in the probability space and the coefficients are necessary to guarantee that the system is solvable, and we proceed next to describe the elements of the FBSDEs (2.1)–(2.3). We begin specifying the class of random environment processes to be considered.

Definition 2.1 Let μ be a measurable map from $(\Omega, \mathcal{F}, \mathbb{P})$ into

$$\Big(\mathrm{D}\big([0,T]; \mathcal{P}_2(\mathbb{R}^d)\big), \mathcal{D}\big([0,T]; \mathcal{P}_2(\mathbb{R}^d)\big)\Big).$$

For each $0 \le t \le T$, denote by μ_t the projection of μ at time t, that is, $\mu_t(\omega) := e_t \circ \mu(\omega)$. We say that μ is a *random environment* if there exists an $\mathcal{F}_0^\mu$-measurable variable X_0 and a probability measure $\nu_0 \in \mathcal{P}_2(\mathbb{R}^d)$, the *initial distribution of* μ, such that

$$\mathbb{P} \circ X_0^{-1} = \nu_0, \qquad \mathbb{P}\,[\mu_0 = \nu_0] = 1. \tag{2.4}$$

We are interested in *Marked Poisson processes with random intensity and marks on a measurable space* $(\mathbf{R}, \mathcal{R})$*, such that the information from the environment influences their behaviour*. This is done assuming, first, that the σ−field $\mathbb{F}^\mu \subset \mathbb{F}$ for some general filtration $\mathbb{F}$. Let $n_0 \ge 1$ be the fixed dimension of the jump process, and $\lambda : \Omega \times [0,T] \times \mathbf{R} \to \mathbb{R}_+^{n_0}$ is an $\mathbb{F}$*-predictable intensity process*. Denoting by $x^{(i)}$ the i-th component of a given multidimensional quantity x, we will implicitly assume throughout that λ satisfies the following conditions:

1. Each coordinate $\lambda_t(r)^{(i)}$ is a non-negative process independent of $\lambda_t(r)^{(j)}$ for all $i \neq j$ with $i, j = 1, 2, \ldots, n_0$, and all $(t, r) \in [0,T] \times \mathbf{R}$.
2. For any $0 \le a < b \le T$,

$$\int_a^b \int_{\mathbf{R}} \lambda_t(r)^{(i)} Q(\mathrm{d}r)^{(i)}\, \mathrm{d}t < \infty \qquad \mathbb{P}-\text{a.s.},$$

where $Q^{(1)}, \ldots, Q^{(n_0)}$ are finite measures on $(\mathbf{R}, \mathcal{R})$.

Then, we say that the random kernel $K^\lambda : \Omega \times [0, T] \times \mathcal{R} \to \mathbb{R}_+^{n_0}$, defined entry-wise as

$$K_t^\lambda(\omega, C)^{(i)} := \int_C \lambda_t(\omega, r)^{(i)} Q(\mathrm{d}r)^{(i)}, \qquad i = 1, \ldots, n_0, \tag{2.5}$$

for all $(\omega, t) \in \Omega \times [0, T]$ and all $C \in \mathcal{R}$, is an $\mathbb{F}$-*intensity kernel.*

Remark 2.1

(i) In the context of point processes, K^λ is simply a multivariate process such that each coordinate is an $\mathbb{F}$-predictable kernel from $\big(\Omega \times [0, T], \mathcal{F} \otimes \mathcal{B}([0, T])\big)$ into $(\mathbf{R}, \mathcal{R})$. That is,

- $K_\cdot(\cdot, C)^{(i)}$ is $\mathbb{F}$-predictable for each $C \in \mathcal{R}$,
- $K_t(\omega, \cdot)^{(i)}$ is a finite measure on $(\mathbf{R}, \mathcal{R})$ for each $(\omega, t) \in \Omega \times [0, T]$;

see (1.25).

(ii) In general, when we refer to (N^λ, ξ) as an n_0-dimensional marked Poisson process, we allude to the family

$$(N^\lambda, \xi) = \left\{(N^{\lambda(i)}, \xi^{(i)}), i = 1, \ldots, n_0\right\},$$

where each $(N^{\lambda(i)}, \xi^{(i)})$ is an independent marked Poisson process with an $\mathbb{F}$-intensity kernel $K^\lambda(\cdot)^{(i)}$ from $(\Omega \times [0, T], \mathcal{F} \otimes \mathcal{B}([0, T]))$ into $(\mathbf{R}, \mathcal{R})$. Similarly, we write K^λ (resp. η^λ, to be defined below) for the vector with i-th entry $K^\lambda(\cdot)^{(i)}$ (resp. $\eta^\lambda(\cdot)^{(i)}$); see (2.5) and (2.6).

To clearly describe the construction of the jump process, we illustrate its elements and some examples for the one-dimensional case. As we have seen, the *jump noise process* is defined as a marked Poisson process (N^λ, ξ) on $[0, T]$ with marks on $\mathbf{R}$, with $\mathbb{F}$-intensity kernel K^λ. From Lemma 1.3, this means that the first component N^λ is a Poisson process with stochastic intensity $K_t^\lambda(\mathbf{R})$ and jumping times $\tau = \{\tau_n, n \geq 1\}$, while the second component, $\xi = \{\xi_n, n \geq 1\}$, is a sequence of $\mathbf{R}$-valued random variables with transition probabilities

$$\mathbb{P}\left[\xi_n \in C | \sigma\left\{(\tau_1, \xi_1), \ldots, (\tau_{n-1}, \xi_{n-1})\right\} \vee \sigma\left\{\tau_n\right\}\right] = \frac{K_{\tau_n}^\lambda(C)}{K_{\tau_n}^\lambda(\mathbf{R})}.$$

Note that (N^λ, ξ) can also be characterized by its so-called *lifting* [20] (Chapter 5); that is, the random measure η^λ defined as

$$\eta^\lambda(B) = \sum_{n \geq 1} \delta_{(\tau_n, \xi_n)}(B), \qquad B \in \mathcal{B}([0, T]) \otimes \mathcal{R}, \tag{2.6}$$

with predictable compensator $K_t^\lambda(\mathrm{d}r)\,\mathrm{d}t$. Consequently, the compensated measure $\tilde{\eta}^\lambda$ is simply defined as

$$\tilde{\eta}^\lambda(\omega, \mathrm{d}t, \mathrm{d}r) := \eta^\lambda(\omega, \mathrm{d}t, \mathrm{d}r) - K_t^\lambda(\omega, \mathrm{d}r)\,\mathrm{d}t. \tag{2.7}$$

Note that this definition immediately allows us to consider well-known point processes, such as the ones listed below, together with their compensated measure $\tilde{\eta}^\lambda$.

Counting Processes Let $\mathbf{R} = \{1\}$ and $Q(\mathbf{R}) = 1$. In this case, η^λ corresponds to the lifting of a point process on the line with jumps of size one ($\xi_n = 1$) at times τ_n. In other words,

$$\eta^\lambda(\omega, \mathrm{d}t, \mathrm{d}\zeta) = \sum_{s \geq 0} \mathbf{1}_{\{\Delta N^\lambda \neq 0\}}(\omega, s)\delta_{(s, \Delta N_s^\lambda(\omega))}(\mathrm{d}t, \mathrm{d}\zeta) = \sum_{n \geq 1} \delta_{(\tau_n, 1)}(\mathrm{d}t, \mathrm{d}\zeta),$$

where $\{\tau_n\}$ are the jump times of a simple counting process N^λ with intensity λ_t. Moreover: from the theory of Poisson random measures we have that

$$\int_{(0,t] \times \mathbf{R}} \eta^\lambda(\mathrm{d}s, \mathrm{d}r) = N_t^\lambda,$$

with the compensated measure corresponding to

$$\int_{(0,t] \times \mathbf{R}} \left\{\eta^\lambda(\mathrm{d}s, \mathrm{d}r) - \lambda_s\,\mathrm{d}s \otimes \delta_{\{1\}}(\mathrm{d}r)\right\} = N_t^\lambda - \int_0^t \lambda_s\,\mathrm{d}s = \tilde{N}_t^\lambda;$$

i.e. $\tilde{N}^\lambda$ is the corresponding compensated $\mathbb{F}$-martingale. The process N^λ can be built explicitly by following a standard time scaling argument from a unit-rate Poisson point process; see [20, Thm. 5.4.1].

From here the type of counting process can be further specified by the choice of λ; for example, Poisson processes are obtained by setting λ_t as a constant (homogeneous Poisson of rate λ) or as a deterministic function (non-homogeneous Poisson). Instances of counting processes with stochastic intensity such as the doubly stochastic Poisson and even some self-exciting processes can be considered here, but will be discussed later.

Markov Renewal Processes Let $\mathbf{R} \subset \mathbb{R}$ be a non-empty finite set and $Q(\mathbf{R}) = 1$. Notice that the choice of a finite space of marks is mainly for illustrative purposes to avoid further conditions on Q at the moment. Let λ be an integrable, non-negative predictable function, and set K^λ to be a kernel of the form

$$K_t^\lambda(\mathrm{d}r) = \lambda_t Q(\mathrm{d}r);$$

i.e. (N^λ, ξ) is a marked Poisson process with intensity λ and i.i.d. marks distributed Q. It is a known fact from the renewal theory that

$$\boldsymbol{\xi}_t := \sum_{n=1}^{N_t^\lambda} \xi_n = \int_{(0,t]\times\mathbf{R}} r\, \eta^\lambda(\mathrm{d}t, \mathrm{d}r), \tag{2.8}$$

with η^λ as in (2.6), defines a renewal-reward process. In fact, by focusing only on the marks,

$$\xi_t := \xi_n \mathbf{1}_{[\tau_n, \tau_{n+1})}(t) = \int_{\mathbf{R}} r\, \eta^\lambda(\{t\}, \mathrm{d}r)$$

produces a semi-Markov process in general, and a Markov chain in continuous time when the inter-renewal times $\{\tau_{n+1} - \tau_n, n \geq 0\}$ are i.i.d. [29, Sec. 10.3]. We will return to the topic of Markov chains later, when discussing regime-switching diffusions.

Pure Jump Lévy Processes Let $\mathbf{R} = \mathbb{R}$, and let $K_t^\lambda(\mathrm{d}r) \equiv Q(\mathrm{d}r)$ with

$$Q(\{0\}) = 0 \qquad \text{and} \qquad \int_{\mathbf{R}} \min\left\{r^2, 1\right\} Q(\mathrm{d}r) < \infty.$$

Then, the process

$$L_t := \int_{(0,t]\times\{|r|\leq 1\}} r\, \tilde{\eta}^\lambda(\mathrm{d}t, \mathrm{d}r) + \int_{(0,t]\times\{|r|>1\}} r\, \eta^\lambda(\mathrm{d}t, \mathrm{d}r),$$

where η^λ is the lifting (as in (2.6)) of a marked Poisson process (N^λ, ξ) with $\lambda \equiv 1$ and i.i.d. marks with law Q, defines a *pure jump Lévy process*. This type of processes are well-known in the literature, including stochastic control theory; see, for instance, [57] and references therein.

We can now describe the probability space endowed with both an idiosyncratic noise and the jump noise process. Hereafter we write d_1 and d_0 to denote the dimension of forward and backward components, respectively; likewise, we use n_1 and n_0 to represent the dimension of the continuous and discontinuous noise inputs, respectively.

Definition 2.2 A tuple $(\Omega, \mathcal{F}, \mathbb{F}, \mathbb{P}, X_0, \mu, K^\lambda, (N^\lambda, \xi), W)$ –or simply $(X_0, \mu, K^\lambda, (N^\lambda, \xi), W)$– is said to be an *admissible set-up* if

1. μ is a random environment on $(\Omega, \mathcal{F}, \mathbb{P})$;
2. $X_0 \in L^2(\Omega, \mathcal{F}_0, \mathbb{P}; \mathbb{R}^d)$;
3. K^λ is a non-negative, $\mathbb{F}$-predictable, locally integrable kernel from $\big(\Omega \times [0, T], \mathcal{F} \otimes \mathcal{B}([0, T])\big)$ into the measurable space $(\mathbf{R}, \mathcal{R})$;

4. (N^λ, ξ) is an n_0-dimensional marked Poisson process, adapted to $\mathbb{F}$, with $\mathbb{F}$-stochastic intensity kernel K^λ and lifting η^λ;
5. W is an n_1–dimensional $\mathbb{F}$–Wiener processes, referred as an *idiosyncratic noise*;
6. (X_0, μ, η^λ) is independent of W (under $\mathbb{P}$);
7. Any $\mathbb{F}^{(X_0,\mu,\eta^\lambda,W)}$-martingale is also an $\mathbb{F}$-martingale (under $\mathbb{P}$).

Remark 2.2

(i) Condition 7 is commonly known in the literature as the *$\mathcal{H}$-hypothesis*, which means that $\mathbb{F}^{(X_0,\mu,\eta^\lambda,W)} \hookrightarrow \mathbb{F}$, or simply that $(X_0, \mu, \eta^\lambda, W)$ is *compatible* with $\mathbb{F}$; see [3].

(ii) Since we only consider kernels K^λ that are obtained through an intensity process λ (see (2.5)), we write $(X_0, \mu, \lambda, (N^\lambda, \xi), W)$ when referring to the admissible set-up $(X_0, \mu, K^\lambda, (N^\lambda, \xi), W)$.

Next, we describe the spaces of integrals with respect to the compensated measure $\tilde{\eta}^\lambda$ under the environment μ:

$$\begin{aligned}\mathfrak{H}^{\lambda,\mu} := \Big\{ U : \Omega \times [0,T] \times \mathbf{R} \to \mathbb{R}^{d_0 \times n_0} \Big| \; & U \text{ is } \mathbb{F} \otimes \mathcal{R} - \text{predictable,} \\ & \text{with } \|U_t(\omega,\cdot)\|_{\lambda_t(\omega)} < \infty \\ & \mathrm{d}\mathbb{P} \otimes \mathrm{d}t\text{–a.e.}(\omega,t) \in \Omega \times [0,T] \Big\},\end{aligned} \tag{2.9}$$

where $\|\cdot\|^2_{\lambda_t(\omega)} = (\cdot\,,\,\cdot)_{\lambda_t(\omega)}$ is a random norm defined for matrix-valued functions, induced by the inner product

$$(u \cdot v)_{\lambda_t(\omega)} := \int_{\mathbf{R}} \mathrm{Tr}\left(u(r)\, \mathrm{diag}\left(K^\lambda(\omega, t; \mathrm{d}r)\right) v(r)^\top \right), \tag{2.10}$$

for $\mathrm{d}\mathbb{P} \otimes \mathrm{d}t$-a.a. $(\omega, t) \in \Omega \times [0, T]$. This is due to the fact that, for each $U \in \mathfrak{H}^{\lambda,\mu}$ (writing $U_t(\cdot) = U(t, \cdot)$), if

$$\int_0^T \|U_t(\cdot)\|^2_{\lambda_t}\, \mathrm{d}t < \infty \qquad \mathbb{P} - \text{a.s.}, \tag{2.11}$$

then the integral process defined below is well-defined $\mathrm{d}\mathbb{P} \otimes \mathrm{d}t$-almost everywhere in $\Omega \times [0, T]$:

$$\int_{(0,t]\times\mathbf{R}} U_s(r)\eta^\lambda(\mathrm{d}s, \mathrm{d}r) := \sum_{j=1}^{n_0} \sum_{n\geq 1} U^{(\cdot,j)}_{\tau_n^{(j)}}(\xi_n^{(j)}) \mathbf{1}_{\{\tau_n^{(j)} \leq t\}}, \tag{2.12}$$

where $\tau^{(j)}$ (resp. $\xi^{(j)}$) is the sequence of jump times (marks) of the j-th member of (N^λ, ξ), and $U^{(\cdot,j)}$ denotes the j-th column of U. Moreover, if the random variable in (2.11) is also integrable with respect to $\mathbb{P}$, then the integrals with respect to the compensated martingale measure $\tilde{\eta}^\lambda$ from (2.7) are also square-integrable martingales, see [20, Thm. 5.1.33-34].

We refer to (2.12) as the *integral of U with respect to* (N^λ, ξ) or, equivalently, the *integral of U with respect to* η^λ. Following the notation used for well-posed integrals with respect to square-integrable martingales, we write

$$\mathbb{H}^2(\tilde{\eta}^\lambda) := \left\{ U \in \mathfrak{H}^\lambda \;\middle|\; \mathbb{E}\left[\int_0^T \|U_t(\cdot)\|^2_{\lambda_t}\, \mathrm{d}t\right] < \infty \right\}$$

for the space of well-posed integrands with respect to (N^λ, ξ), and

$$\mathcal{L}^2(\tilde{\eta}^\lambda) := \left\{ \left(\int_{(0,t]\times\mathbf{R}} U_s(r)\tilde{\eta}^\lambda(\mathrm{d}s, \mathrm{d}r) \right)_{t\in[0,T]} \;\middle|\; U \in \mathbb{H}^2(\tilde{\eta}^\lambda) \right\}$$

for the space of martingales that can be expressed as compensated integrals with respect to (N^λ, ξ).

Remark 2.3

(i) For the rest of this chapter we shall omit the environment dependence of $\mathfrak{H}^{\lambda,\mu}$ and simply write $\mathfrak{H}^\lambda$ to alleviate some of the notation. This should not create confusion as the environment remains fixed, as opposed to some of the results in later chapters.

(ii) Observe that for the unmarked case, $(u \cdot v)_{\lambda_t} = \mathrm{Tr}\left(u \,\mathrm{diag}\,(\lambda_t)\, v^\top\right)$ and

$$\int_0^t U_s\, \mathrm{d}\tilde{N}^\lambda_s := \sum_{j=1}^{n_0} \left\{ \sum_{\tau_n \le t} U^{(\cdot,j)}_{\tau_n^{(j)}} - \int_0^t U_s^{(\cdot,j)} \lambda_s^{(j)}\, \mathrm{d}s \right\},$$

for all $t \in [0, T]$.

Note that solving the backward component (2.2) in the presence of a random environment requires taking into account the additional information contained in $\mathbb{F}$, in addition to the generated by the noise processes, since its solution cannot depend on future information of the process μ. However, the nature of the noise we consider admits the possibility of applying the existing theory of *backward stochastic differential equations* (BSDEs). When the random environment μ is not present in the dynamics and the forward component is absent, the solution of the BSDE system was tackled by applying well-known martingale representation results; see, for instance, the seminal work of Peng, El-Karoui, Quenez and Sulem on this subject [34, 61, 66]. The idea behind this approach consists of including additional processes as part of the solution to BSDEs that modulate the interactions with the components of the set-up; for example, in the classical case where only a Brownian motion is present, the solution to a BSDE consists of a pair (Y, Z) such that Y is an Itô process with a specified terminal condition and Z is the corresponding diffusion term.

In general, we have the following: Let $\tilde{M} \in \mathcal{H}^2(\mathbb{R}^{d_0})$ be an $\mathbb{F}$-adapted martingale defined on $(\Omega, \mathcal{F}, \mathbb{P})$, admiting a decomposition of the form $\tilde{M} = \tilde{M}_0 + \tilde{M}^c + \tilde{M}^d$, with $\tilde{M}_0$ an $\mathcal{F}_0$-measurable random variable, $\tilde{M}^c$ a square-integrable martingale with continuous paths and $\tilde{M}^d$ a purely discontinuous, square integrable martingale. Intuitively, by martingale representation theorems, $\tilde{M}^c$ and $\tilde{M}^d$ are related to W and (N^λ, ξ), respectively. However, in our case, these processes do not represent all the randomness affecting the system, and therefore, in order to ensure that martingale $\tilde{M}$ is $\mathcal{F}_t$-measurable at each time t, an additional term must be added.

Remark 2.4 In order to address the difficulties discussed above, we now present a couple of useful results from the theory of BSDEs; see Propositions 2.5–2.6 and Corollary 2.7 in [58]. The original results were presented in a general form, however, for the sake of completeness and readability, we reproduce them here in the form of a single proposition, adapted to our framework and notation, restricting ourselves to the case of marked Poisson point processes.

Proposition 2.1 *Each random variable $\zeta \in L^2(\mathcal{F}_T; \mathbb{R}^{d_0})$ has an orthogonal decomposition with respect to (N^λ, ξ) and W. Namely, there exists a unique (up to $\mathbb{P}$-indistinguishability) triplet (Z, U, M) such that:*

1. *$Z \in \mathbb{H}^2(W)$ and $U \in \mathbb{H}^2(\tilde{\eta}^\lambda)$ of dimensions $d_0 \times n$ and $d_0 \times n_0$, respectively;*
2. *the stochastic processes:*

$$\left(\int_0^t Z_s \, \mathrm{d} W_s, t \in [0, T]\right), \quad \left(\int_{(0,t]\times\mathbf{R}} U_s(r)\tilde{\eta}^\lambda(\mathrm{d} s, \mathrm{d} r), t \in [0, T]\right);$$

   are $\mathbb{R}^{d_0}$-valued, $\mathbb{F}$-adapted and independent martingales;
3. *$M \in \mathcal{H}^2(\mathbb{R}^d)$ with $M_0 = 0$ and orthogonal to W and $\tilde{\eta}^\lambda$, i.e. $\langle M, W\rangle = 0$ and*

$$\widetilde{\Delta M} := \frac{\mathbb{E}\left[\Delta M \eta^\lambda(\mathrm{d} s, \mathrm{d} r)\right] |_{\tilde{\mathbb{F}}\otimes\mathcal{R}}}{\mathbb{E}\left[\eta^\lambda(\mathrm{d} s, \mathrm{d} r)\right] |_{\tilde{\mathbb{F}}\otimes\mathcal{R}}} = 0, \tag{2.13}$$

   where ΔM denotes the jumps of the process M, i.e. $\Delta M_t := M_t - M_{t-}$.
4. *the random variable ζ can be decomposed as:*

$$\zeta = \mathbb{E}[\zeta | \mathcal{F}_0] + \int_0^T Z_s \, \mathrm{d} W_s + \int_{[0,T]\times\mathbf{R}} U_s(r)\tilde{\eta}^\lambda(\mathrm{d} s, \mathrm{d} r) + M_T.$$

Remark 2.5 The term at the l.h.s. of (2.13) corresponds to the $\mathbb{F}$-predictable projection of ΔM on η^λ; see Definition 2.1 at [58]. It is a known fact from the theory of stochastic integration that under suitable conditions (integer-valued integrable random measures, see Chapter III of [42] and Section 13.3 of [28]), it determines the quadratic variation of integrals with respect to a random measure. In our particular case, Eq. (2.13) implies that

$$\left\langle M, \int_{[0,\cdot]\times\mathbf{R}} U_t(r)\tilde{\eta}^\lambda(\mathrm{d}t,\mathrm{d}r)\right\rangle = \int_{[0,T]\times\mathbf{R}} \widetilde{\Delta M U_t}(r)K_t^\lambda(\mathrm{d}r)\,\mathrm{d}t = 0$$

for all $U \in \mathbb{H}^2(\tilde{\eta}^\lambda)$, which is consistent with the usual notion of orthogonality.

2.2 Comparison with Uncoupled Equations

When the forward and backward components of the system of interest are decoupled, their existence and uniqueness can be obtained from the existing literature, analyzing independently each component, under a set of measurability and Lipschitz conditions on the coefficients. Later, it will be seen that these conditions are consistent with the coupled forward-backward case analyzed here.

More precisely, given the random environment μ *(fixed)*, define the *SDE with coefficients* (b, σ, γ) *and initial condition* X_0, as the solution of the integral equation

$$\begin{aligned} X_t =& X_0 + \int_0^t b(s,\mu_s,X_s)\,\mathrm{d}s + \int_0^t \sigma(s,\mu_s,X_s)\,\mathrm{d}W_s \\ &+ \int_{(0,t]\times\mathbf{R}} \gamma(s,r,\mu_{s-},X_{s-})\tilde{\eta}^\lambda(\mathrm{d}s,\mathrm{d}r), \end{aligned} \tag{2.14}$$

for $t \in [0, T]$, where

$$\begin{aligned} (b,\sigma) &: \Omega\times[0,T]\times\mathcal{P}_2(\mathbb{R}^d)\times\mathbb{R}^d \to \mathbb{R}^d\times\mathbb{R}^{d\times n_1}, \\ \gamma &: \Omega\times[0,T]\times\mathbf{R}\times\mathcal{P}_2(\mathbb{R}^d)\times\mathbb{R}^d \to \mathbb{R}^{d\times n_0}. \end{aligned} \tag{2.15}$$

Here, the r.h.s. of (2.14) should be understood as a component-wise $\mathbb{P}$-a.s. equality

$$X_0^{(i)} + \int_0^t b_s^\mu(X)^{(i)}\,\mathrm{d}s + \int_0^t \sigma_s^\mu(X)^{(i,\cdot)}\,\mathrm{d}W_s + \int_{(0,t]\times\mathbf{R}} \gamma_s^\mu(r,X)^{(i,\cdot)}\tilde{\eta}^\lambda(\mathrm{d}s,\mathrm{d}r),$$

for each $i = 1, \ldots, d$, where, to ease the notation, we have written the dependence on the time and environment as

$$\left.\begin{aligned} (b,\sigma)_t^\mu(X)(\omega) &:= (b,\sigma)(\omega,t,\mu_t(\omega),X_t(\omega)), \\ \gamma_t^\mu(r,X)(\omega) &:= \gamma(\omega,t,r,\mu_{t-}(\omega),X_{t-}(\omega)), \end{aligned}\right\} \forall(\omega,t)\in\Omega\times[0,T],$$

as well as the matrix notation $A(\cdot)^{(i,\cdot)}$ for the i-th row of a finite dimensional matrix-valued function $A(\cdot)$. Throughout the book we will extend this short hand matrix-notation to $A(\cdot)^{(\cdot,j)}$ and $A(\cdot)^{(i,j)}$ when referring to the j-th column and (i, j)-th entry of $A(\cdot)$, respectively.

In the rest of this section we assume that the coefficients of the SDE (BSDE) are $\mathbb{F}$-predictable and stochastic Lipschitz, even without an explicit mention; see Condition (B.4) below, Section V.3 at [65] or condition **(F3)** at [58]. Later we shall see that these correspond to simplified versions of the assumptions established in the main result of this chapter, Theorem 2.1. Next result follows from Theorems 14.21-14.23 in [42].

Proposition 2.2 *Let* (b, σ, γ) *be defined as in* (2.15). *If* (b, σ, γ) *are of at most linear growth on* x, *then there exists a unique* $\mathbb{F}$*-adapted process* X, *defined on* $(\Omega, \mathcal{F}, \mathbb{P})$ *with* $\|X\|_* < \infty$, *solving the SDE* (2.14).

Similarly, for the backward component in a random environment, Proposition 2.1 provides the first elements to write BSDEs with a terminal condition $\zeta \in L^2(\mathcal{F}_T; \mathbb{R}^{d_0})$, defined on an admissible set-up, in terms of an additional martingale M. Namely,

$$Y_t = \zeta + \int_t^T f_s \,\mathrm{d}s - \int_t^T Z_s \,\mathrm{d}W_s - \int_{(t,T]\times\mathbf{R}} U_s(r)\tilde{\eta}^\lambda(\mathrm{d}s, \mathrm{d}r) - (M_T - M_t), \tag{2.16}$$

where f is the driver of the system, an $\mathbb{F}$-predictable process. When f has the form

$$f_t(\omega) = f(\omega, t, \mu_t(\omega), Y_t(\omega), Z_t(\omega), U_t(\omega, \cdot)),$$

we say that (2.16) is a *BSDE on the random environment* μ .

Remark 2.6 Just as in the forward case, (2.16) should be understood as a handy matrix notation for the component-wise identity

$$\begin{aligned} Y_t^{(i)} =& \zeta^{(i)} + \int_t^T f_s^{(i)} \,\mathrm{d}s - \int_t^T Z_s^{(i,\cdot)} \,\mathrm{d}W_s - \int_{(t,T]\times\mathbf{R}} U_s(r)^{(i,\cdot)}\tilde{\eta}^\lambda(\mathrm{d}s, \mathrm{d}r^j) \\ &- (M_T^{(i)} - M_t^{(i)}). \end{aligned}$$

As before, this particular form of driver f, and the choice made for noise process, allows us to rely on existent results of the general theory of BSDEs. Since the predictable quadratic variation of both W and $\tilde{\eta}^\lambda$ is absolutely continuous with respect to the Lebesgue measure on $[0, T]$, and an orthogonal martingale decomposition holds, the existence of a unique solution process (Y, Z, U, M) can be obtained.

The next result follows from Theorem 3.5 and Corollary 3.6 in [58], under the following hypothesis on the driver.

Assumptions A *Let*

$$f : \Omega \times [0, T] \times \mathcal{P}_2(\mathbb{R}^d) \times \mathbb{R}^{d_0} \times \mathbb{R}^{d_0 \times n_1} \times \mathfrak{H}^\lambda \to \mathbb{R}^{d_0}$$

be a driver, and let $\underline{K}$, $\underline{K}^W$ and $\underline{K}^{\eta^\lambda}$ the $\mathbb{F}$-predictable Lipschitz coefficients w.r.t. y, z and u, respectively. Let

$$\alpha_t^2 := \max\left\{\sqrt{\underline{K}_t}, \underline{K}_t^W, \underline{K}_t^{\eta^\lambda}\right\}, \qquad \overline{\alpha}_t := \int_0^t \alpha_s^2 \, \mathrm{d}s. \tag{2.17}$$

Moreover, the following conditions hold for some sufficiently large $\beta^* > 0$:

$$\mathbb{E}\left[e^{\beta^*\overline{\alpha}_T} |\zeta|^2\right] < \infty \qquad \textit{and} \qquad \mathbb{E}\left[\int_0^T e^{\beta^*\overline{\alpha}_t} \frac{|f_t^\mu(0,0,0)|}{\alpha_t^2} \mathrm{d}t\right] < \infty.$$

Proposition 2.3 *Under Assumptions A, there exists a unique (up to $\mathbb{P}$-indistinguishability) process (Y, Z, U, M) with*

$$Y : \Omega \times [0,T] \to \mathbb{R}^{d_0}, \qquad Z : \Omega \times [0,T] \to \mathbb{R}^{d_0 \times n_1},$$
$$U : \Omega \times [0,T] \times \mathbf{R} \to \mathbb{R}^{d_0 \times n_0}, \qquad M : \Omega \times [0,T] \to \mathbb{R}^{d_0},$$

such that

- *Y and M are $\mathbb{F}$-adapted, Z is $\mathbb{F}$-progressively measurable, and U is $\mathbb{F} \otimes \mathcal{R}$-predictable;*
- *(Y, Z, U, M) is a solution process to the BSDE in random environment* (2.16)*;*
- $\|(Y,Z,U,M)\|_{*,\beta^*,\alpha}^2 := \|Y\|_{\mathcal{S}_{\beta,\alpha}^2}^2 + \|Z\|_{\mathbb{H}_{\beta,\alpha}^{2,W}}^2 + \|U\|_{\mathbb{H}_{\beta,\alpha}^{2,\tilde{\eta}^\lambda}}^2 + \|M\|_{\mathcal{H}_{\beta,\alpha}^{2,\perp}}^2 < \infty,$

  where

$$\|Y\|_{\mathcal{S}_{\beta,\alpha}^2}^2 := \mathbb{E}\left[\sup_{t\in[0,T]}\left\{e^{\beta^*\overline{\alpha}_t}|Y_t|^2\right\}\right],$$

$$\|Z\|_{\mathbb{H}_{\beta,\alpha}^{2,W}}^2 := \mathbb{E}\left[\int_0^T e^{\beta^*\overline{\alpha}_t}\|Z_t\|_{\mathrm{Fr}}^2\,\mathrm{d}t\right],$$

$$\|U\|_{\mathbb{H}_{\beta,\alpha}^{2,\tilde{\eta}^\lambda}}^2 := \mathbb{E}\left[\int_0^T e^{\beta^*\overline{\alpha}_t}\|U_t(\cdot)\|_{\lambda_t}^2\,\mathrm{d}t\right],$$

$$\|M\|_{\mathcal{H}_{\beta,\alpha}^{2,\perp}}^2 := \mathbb{E}\left[\int_0^T e^{\beta^*\overline{\alpha}_t}\,\mathrm{d}\,\mathrm{tr}\,\langle M\rangle_t\right].$$

Henceforth, we write

$$\mathcal{S}_{\beta^*,\alpha}^2 \times \mathbb{H}_{\beta^*,\alpha}^{2,W} \times \mathbb{H}_{\beta^*,\alpha}^{2,\tilde{\eta}^\lambda} \times \mathcal{H}_{\beta^*,\alpha}^{2,\perp}$$
$$:= \Big\{(Y,Z,U,M) \in \mathcal{S}^2(\mathbb{R}^{d_0}) \times \mathbb{H}^2(W) \times \mathbb{H}^2(\tilde{\eta}^\lambda) \times \mathcal{H}^{2,\perp}$$
$$\text{such that } \|(Y,Z,U,M)\|_{*,\beta^*,\alpha}^2 < \infty\Big\}.$$

Remark 2.7 Note that, based on the previous results for uncoupled systems, the following linear FBSDEs have a unique (strong) solution (X, Y, Z, U, M) defined in the space $\mathcal{S}^2(\mathbb{R}^d)\times\mathcal{S}^2(\mathbb{R}^{d_0})\times\mathbb{H}^2(W)\times\mathbb{H}^2(\tilde{\eta}^\lambda)\times\mathcal{H}^{2,\perp}$, under suitable conditions on their coefficients. See, for instance, [28, 34, 74], where explicit solutions can be found in some particular cases, as in the case when no jump term is included:

1. Uncoupled (linear) forward component:

$$X_t = X_0 + \int_0^t B_s \,\mathrm{d}s + \int_0^t \Sigma_s \,\mathrm{d}W_s + \int_{(0,t]\times\mathbf{R}} \Gamma_s(r)\tilde{\eta}^\lambda(\mathrm{d}s,\mathrm{d}r), \tag{2.18}$$

$$\begin{aligned} Y_t =& (GX_T + \zeta) + \int_t^T (GX_s + F_s)\,\mathrm{d}s - \int_t^T Z_s \,\mathrm{d}W_s \\ &- \int_{(t,T]\times\mathbf{R}} U_s(r)\tilde{\eta}^\lambda(\mathrm{d}s,\mathrm{d}r) - (M_T - M_t). \end{aligned} \tag{2.19}$$

2. Uncoupled (linear) backward component:

$$\begin{aligned} X_t =& X_0 + \int_0^t \left(B_s + G^\top Y_s\right)\mathrm{d}s + \int_0^t \left(\Sigma_s + G^\top Z_s\right)\mathrm{d}W_s \\ &+ \int_{(0,t]\times\mathbf{R}} \left(\Gamma_s(r) + G^\top U_s(r)\right)\tilde{\eta}^\lambda(\mathrm{d}s,\mathrm{d}r), \end{aligned} \tag{2.20}$$

$$\begin{aligned} Y_t =& \zeta + \int_t^T F_s \,\mathrm{d}s - \int_t^T Z_s \,\mathrm{d}W_s - \int_{(t,T]\times\mathbb{R}_+} U_s(r)\tilde{\eta}^\lambda(\mathrm{d}s,\mathrm{d}r) \\ &- (M_T - M_t). \end{aligned} \tag{2.21}$$

In the next section these type of equations will be useful to prove the existence of solution in the general case (2.1) and (2.2), implementing a linear perturbation to the convergence iterative method.

2.3 Existence and Uniqueness of Solutions

The following is the base set of assumptions that is maintained throughout the entire book to guarantee the existence of an admissible set-up, according to Definition 2.2. These results require a particular structure of the probability space, which is basically composed of two elements, described below.

2.3.1 Probability Space and Assumptions

Let Ω^0 and Ω^1 be complete and separable metric spaces, endowed with Borel σ-algebras $\mathcal{F}^0$ and $\mathcal{F}^1$, respectively. We assume that $(\Omega^0, \mathcal{F}^0, \mathbb{P}^0)$ is a complete probability space supporting an n_0-dimensional Poisson process N. Similarly, $(\Omega^1, \mathcal{F}^1, \mathbb{P}^1)$ is a complete probability space supporting an n_1-dimensional Wiener process W. Here, we adopt the usual approach to define an n_0-dimensional Poisson process, which is given as a vector $N = (N^{(1)}, \ldots, N^{(n_0)})^\top$, where each entry is an independent copy from the rest, $Q^{(i)}$ is a probability measure on $(\mathbf{R}, \mathcal{R})$, and each coordinate

$$N((0,\cdot] \times C \times D)^{(i)} - \int_{(0,\cdot]\times C\times D} \mathrm{d}t \otimes Q(\mathrm{d}r)^{(i)} \otimes \mathrm{d}s$$

is an $\mathbb{F}^N$-martingale, for all $C \in \mathcal{R}$ and $D \in \mathcal{B}(\mathbb{R}_+)$ such that

$$\int_{(0,t_0]\times C\times D} \mathrm{d}t \otimes Q(\mathrm{d}r)^{(i)} \otimes \mathrm{d}s < \infty, \qquad \forall t_0 \in [0, T],$$

and $N(\{0\} \times \mathbf{R} \times \mathbb{R}) = 0$ with $\mathbb{P}^0$–probability one. Moreover, we assume throughout that there exists a measurable mapping $X_0 : (\Omega^0, \mathcal{F}^0, \mathbb{P}^0) \to (\mathbb{R}^d, \mathcal{B}(\mathbb{R}^d))$ with law $\nu_0 \in \mathcal{P}_2(\mathbb{R}^d)$; i.e.

$$\mathbb{E}^0\left[|X_0|^2\right] = \int_{\mathbb{R}^d} |x|^2\, \nu_0(\mathrm{d}x) < \infty.$$

The probability space we consider hereafter is the completion of the product space

$$(\Omega, \mathcal{F}, \mathbb{F}, \mathbb{P}) = (\Omega^0 \times \Omega^1, \mathcal{F}^0 \otimes \mathcal{F}^1, \mathbb{F}^{(X_0,\mu,N)} \otimes \mathbb{F}^W, \mathbb{P}^0 \otimes \mathbb{P}^1). \tag{2.22}$$

The interested reader can find details on the construction of this particular probability space in Theorem 3.1 of [39]. We make the following assumptions on the coefficients of the model described by the FBSDE.

Assumptions B

(B.1) For all $(\nu, x, y, z, u) \in \mathcal{P}_2(\mathbb{R}^d) \times \mathbb{R}^d \times \mathbb{R}^{d_0} \times \mathbb{R}^{d_0\times n_1} \times \mathfrak{H}^\lambda$, the process

$$(b, f, \sigma)(\cdot, \nu, x, y, z, u)$$

take values in $\mathbb{R}^d \times \mathbb{R}^{d_0} \times \mathbb{R}^{d_0\times n_1}$, is $\mathbb{F}$-progressively measurable, and

$$(b, f, \sigma)(\cdot, \mu_\cdot, 0, 0, 0, 0) \in \mathbb{H}^2(W, \tilde{\eta}^\lambda) \times \mathbb{H}^2(W, \tilde{\eta}^\lambda) \times \mathbb{H}^2(W)$$

(B.2) For each $(\nu, x, y, z, u) \in \mathcal{P}_2(\mathbb{R}^d) \times \mathbb{R}^d \times \mathbb{R}^{d_0} \times \mathbb{R}^{d_0\times n_1} \times \mathfrak{H}^\lambda$, the process

$$\gamma(\cdot, \nu, x, y, z, u)$$

takes values in $\mathbb{R}^{d_0 \times n_0}$, *is* $\mathbb{F} \otimes \mathcal{R}$*-predictable, and*

$$\gamma(\cdot, \mu_\cdot, 0, 0, 0, 0) \in \mathbb{H}^2(\tilde{\eta}^\lambda);$$

(B.3) $g(\nu, x) \in L^2(\mathcal{F}_T)$ *for all* $(\nu, x) \in \mathcal{P}_2(\mathbb{R}^d) \times \mathbb{R}^d$;

(B.4) There exist $\mathbb{F}$*-predictable processes*

$$K, \underline{K}, \underline{K}^W, \underline{K}^{\eta^\lambda} : \Omega \times \mathbb{R}_+ \to \mathbb{R}_+, \tag{2.23}$$

and a positive constant $K^0 > 0$, *such that for any* $x, x' \in \mathbb{R}^d$, $\nu, \nu' \in \mathcal{P}_2(\mathbb{R}^d)$, $y, y' \in \mathbb{R}^{d_0}$, $z, z' \in \mathbb{R}^{d_0 \times n_1}$ *and* $u, u' \in \mathfrak{H}^\lambda$, *the following inequality holds* $\mathrm{d}\mathbb{P} \otimes \mathrm{d}t$*-almost everywhere in* $\Omega \times [0, T]$:

$$\begin{aligned} &|\delta(b, f)_t(\nu, x, y, z, u_t)|^2 + \|\delta\sigma_t(\nu, x, y, z, u_t)\|_{\mathrm{Fr}}^2 + \|\delta\gamma_t(\cdot, \nu, x, y, z, u_t)\|_{\lambda_t}^2 \\ &\quad + |\delta g(\nu, x)|^2 \\ &\le K^0\, \mathcal{W}_2(\nu, \nu')^2 + K_t\, |\delta x|^2 + \underline{K}_t\, |\delta y|^2 + \underline{K}_t^W\, \|\delta z\|_{\mathrm{Fr}}^2 + \underline{K}_t^{\eta^\lambda}\, \|\delta u_t(\cdot)\|_{\lambda_t}^2 . \end{aligned}$$

(B.5) Furthermore, the Lipschitz coefficients $K, \underline{K}, \underline{K}^W, \underline{K}^{\eta^\lambda}$ *from* (2.23) *are* $\mathrm{d}\mathbb{P} \otimes \mathrm{d}t$*-a.e. bounded from below away from zero by a positive constant* $K_* > 0$, *and from above by another positive constant* K^*.

(B.6) There exists a full rank $\mathbb{R}^{d_0 \times d}$ *matrix G such that, for* $(\omega, t) \in \Omega \times [0, T]$ *fixed, the mapping* $\mathbf{G}^\mu : \mathbb{R}^d \times \mathbb{R}^{d_0} \times \mathbb{R}^{d_0 \times n_1} \times \mathfrak{H}^\lambda \to \mathbb{R}^d \times \mathbb{R}^{d_0} \times \mathbb{R}^{d_0 \times n_1} \times \mathfrak{H}^\lambda$, *defined as*

$$\mathbf{G}_t^\mu(x, y, z, u)(\omega) := \begin{bmatrix} -G^\top f_t^\mu(x, y, z, u_t)(\omega) \\ G b_t^\mu(x, y, z, u_t)(\omega) \\ G\sigma_t^\mu(x, y, z, u_t)(\omega) \\ G\gamma_t^\mu(\cdot, x, y, z, u_t)(\omega) \end{bmatrix}, \tag{2.24}$$

exists and satisfies the G-monotonicity condition described next. Namely, there exist nonnegative constants $\beta_1, \beta_2, \beta_3$ *with* $\beta_1 + \beta_2 > 0$ *and* $\beta_2 + \beta_3 > 0$ *such that the following inequalities hold* $\mathrm{d}\mathbb{P} \otimes \mathrm{d}t$*-a.e.:*

$$\begin{aligned} &\Big(\delta\mathbf{G}_t^\mu(x, y, z, u_t) \cdot \delta(x, y, z, u_t)\Big) \\ &\qquad \le -\beta_1\, |G(\delta x)|^2 - \beta_2 \left\{ \left|G^\top(\delta y)\right|^2 + \left\|G^\top(\delta z)\right\|_{\mathrm{Fr}}^2 + \left\|G^\top(\delta u_t)\right\|_{\lambda_t}^2 \right\}, \end{aligned}$$

where inner product at the left hand side corresponds to the sum of inner products in the corresponding (matrix) spaces, and

$$\left(\delta g_T^{\mu}(x) \cdot G(\delta x)\right) \geq \beta_3 \left|G(\delta x)\right|^2,$$

for all $(x, y, z, u), (x', y', z', u') \in \mathbb{R}^d \times \mathbb{R}^{d_0} \times \mathbb{R}^{d_0 \times n_1} \times \mathfrak{H}^{\lambda}$.

Some Remarks on the Assumptions

(a) There are some particular cases for which the matrix G can be described easily; for instance, when $d = d_1$, we can always take $G = \mathrm{Id}_d$. On the other hand, when the coefficients are constant, Assumption (B.6) is not needed to solve the system.

(b) The choice of the G-monotonicity condition is based on ideas adapted from equilibrium problems; see [55, 56]. These problems arise naturally in the fields of optimization, Nash equilibria of non-cooperative games and variational inequalities. In the context of FBSDEs, this condition has been also useful to study fully coupled systems of stochastic differential equations with both initial and terminal conditions; see, for instance, [62, 79]. The implementation of the *method of continuation* for solving the system of FBSDEs adapted in this work depends heavily on this condition.

(c) The conditions satisfied by G, and hence by β_1, β_2 and β_3 in Assumption (B.6) result in one of two cases: $\beta_1, \beta_3 > 0$ with $\beta_2 \geq 0$, or $\beta_2 > 0$ with $\beta_1, \beta_3 \geq 0$. These cases will later determine the shape of the auxiliary systems used in the continuation argument. However, it should be noted that there is not a general a priori condition (e.g. the dimension of the components) to determine the positivity of the β_i's. Indeed: in (2.18) and (2.19) (resp. (2.20) and (2.21)), it is possible to take G independent of μ, with $\beta_1 = \beta_3 = 1$ and $\beta_2 = 0$ (resp. $\beta_1 = \beta_3 = 0$ and $\beta_2 = 1$), independently of the dimension of the state variables.

(d) In later chapters we will see that the operator $\mathbf{G}^{\mu}$ is closely related to the Hamiltonian H, in the context of stochastic optimal control problems. Namely, within the context of an exogenous random environment μ, the existence and uniqueness of solution an FBSDE has to be solved in order for the maximum principle to hold. Particularly, in the scalar case without marks the process $\mathbf{G}^{\mu}$ will coincide with the gradient of H, i.e. $\mathbf{G}^{\mu} = \mathrm{D}\, H^{\mu\top}$, with $G = 1$.

As it was pointed out before, the conditions adopted above are perfectly compatible with the existent results for uncoupled components, making the simplifications that this case implies.

We are now ready to present the main result of the chapter, which proof requires some technical results presented below.

Theorem 2.1 *There exists a unique process*

$$(X, Y, Z, U, M) \in \mathcal{S}^2(\mathbb{R}^d) \times \mathcal{S}^2(\mathbb{R}^{d_0}) \times \mathbb{H}^2(W) \times \mathbb{H}^2(\tilde{\eta}^{\lambda}) \times \mathcal{H}^2(W, \tilde{\eta}^{\lambda})^{\perp}$$

(up to $\mathbb{P}$*-indistinguishability) satisfying the FBSDE* (2.1)–(2.3).

As it was stated before, we usually are interested in the case when the influence of one of the forward or backward components is eliminated from the evolution of the other. Within this framework, the previous result can be simplified in the following way, extending existence and uniqueness of solutions for these systems when their evolution is influenced by a random environment.

Corollary 2.1

(i) Let (f, g) be identically zero; then, the SDE (2.14) *has a unique $\mathbb{F}$-adapted solution X in $\mathcal{S}^2(\mathbb{R}^d)$.*

(ii) Let (b, σ, γ) be identically zero with $g \equiv \zeta \in L^2(\mathcal{F}_T; \mathbb{R}^{d_0})$; then, the BSDE (2.16) *has a unique $\mathbb{F}$-adapted solution (Y, Z, U, M) in $\mathcal{S}^2(\mathbb{R}^{d_0}) \times \mathbb{H}^2(W) \times \mathbb{H}^2(\tilde{\eta}^\lambda) \times \mathcal{H}^{2,\perp}$. Furthermore, the norms $\|\cdot\|_*$ and $\|\cdot\|_{*,\beta,\overline{\alpha}}$ are equivalent for each $\beta \geq 0$.*

Proof For the proof of part (i), observe that by taking a null backward component, the uniform bound of the Lipschitz conditions in Assumption (B.5) imply linear growth, thus satisfying the conditions of Proposition 2.2.

For the proof of part (ii), recall the definitions of α^2 and $\overline{\alpha}$ from (2.17). Observe that

$$e^{\beta K_* T} \|\zeta\|_{L^2} \leq \|\zeta\|_{L^2_{\beta,\alpha}} \leq e^{\beta K^* T} \|\zeta\|_{L^2} < \infty,$$

and

$$\begin{aligned}\frac{e^{\beta K_* T}}{K^*} \mathbb{E}\left[\int_0^T \left|f_t^\mu(0,0,0)\right| \mathrm{d}t\right] &\leq \mathbb{E}\left[\int_0^T e^{\beta\alpha_t} \frac{\left|f_t^\mu(0,0,0)\right|}{\alpha_t^2} \mathrm{d}t\right] \\ &\leq \frac{e^{\beta K^* T}}{K_*} \mathbb{E}\left[\int_0^T \left|f_t^\mu(0,0,0)\right| \mathrm{d}t\right] < \infty\end{aligned}$$

$\mathrm{d}\mathbb{P} \otimes \mathrm{d}t$-a.e. in $\Omega \times [0, T]$, for all $\beta \geq 0$. From Proposition 2.3, by taking β^* sufficiently large, there exists a unique

$$(Y, Z, U, M) \in \mathcal{S}^2_{\beta^*,\alpha} \times \mathbb{H}^{2,W}_{\beta^*,\alpha} \times \mathbb{H}^{2,\tilde{\eta}^\lambda}_{\beta^*,\alpha} \times \mathcal{H}^{2,\perp}_{\beta^*,\alpha}$$

satisfying (2.16). The equivalence between norms is due to the bounds of the process $\overline{\alpha}$,

$$0 < K_* t \leq \overline{\alpha}_t \leq K^* T, \qquad \mathrm{d}\mathbb{P} \otimes \mathrm{d}t - \text{a.e.},$$

and triangule inequality. □

2.3.2 Preliminary Results

We begin by presenting a series of lemmas that will be useful in proving the main theorem. Hereafter we assume that

$$(B, F, \Sigma, \Gamma, \zeta) \in \mathbb{H}^2(W, \tilde{\eta}^\lambda) \times \mathbb{H}^2(W, \tilde{\eta}^\lambda) \times \mathbb{H}^2(W) \times \mathbb{H}^2(\tilde{\eta}^\lambda) \times L^2(\mathcal{F}_T; \mathbb{R}^{d_o}) \tag{2.25}$$

represent arbitrary processes in the corresponding spaces, unless specified otherwise.

The following a priori estimate that we cite, obtained by applying [58, Lemma 3.4 and Proposition 3.13] to our framework, will be useful when tackling the existence problem. For the proof of this result we refer the reader to [39, Lemma 4.7].

Lemma 2.1 *Under Assumption A, let (Y, Z, U, M) and (Y', Z', U', M') be the corresponding solutions to* (2.16) *with driver f^1 and f^2, and terminal conditions ζ and ζ' in $L^2(\mathcal{F}_T; \mathbb{R}^{d_o})$, respectively. Then, there exists $\beta^* > 0$ such that the estimate*

$$\begin{aligned}\|\delta(Y, Z, U, M)\|_*^2 \leq & c_*(f^1, f^2, \beta, T)\, \|\delta\zeta\|_{L^2}^2 \\ & + c_{**}(f^1, f^2, \beta, T)\mathbb{E}\left[\int_0^T \left|\delta f_t^\mu(Y', Z', U')\right| \mathrm{d}t\right],\end{aligned}$$

holds for all $\beta \geq \beta^*$, where $c_*(f^1, f^2, \beta, T)$ and $c_{**}(f^1, f^2, \beta, T)$ are positive constants, and $\delta f^\mu(Y', Z', U') := f^{1\mu}(Y', Z', U') - f^{2\mu}(Y', Z', U')$.

Another helpful tool for our calculations consists in the identity expressed in the next lemma, which can be obtained by applying Itô's formula and using the orthogonality of M with respect to W and $\tilde{\eta}^\lambda$.

Lemma 2.2 *Let $X \in \mathcal{S}^2(\mathbb{R}^d)$ and $(Y, Z, U, M) \in \mathcal{S}^2(\mathbb{R}^{d_o}) \times \mathbb{H}^2(W) \times \mathbb{H}^2(\tilde{\eta}^\lambda) \times \mathcal{H}^{2,\perp}$ be defined processes such that*

$$X_t = X_0 + \int_0^t B_s \,\mathrm{d}s + \int_0^t \Sigma_s \,\mathrm{d}W_s + \int_{[0,t]\times\mathbf{R}} \Gamma_s(r)\tilde{\eta}^\lambda(\mathrm{d}s, \mathrm{d}r),$$

$$Y_t = \zeta + \int_t^T F_s \,\mathrm{d}s - \int_t^T Z_s \,\mathrm{d}W_s - \int_{(t,T]\times\mathbf{R}} U_s(r)\tilde{\eta}^\lambda(\mathrm{d}s, \mathrm{d}r) - (M_T - M_t).$$

Then,

$$\begin{aligned}&\mathbb{E}\left[(X_T \cdot \xi)_G\right] - \mathbb{E}\left[(X_0 \cdot Y_0)_G\right] \\ &= \mathbb{E}\left[\int_0^T \left\{-(X_t \cdot F_t)_G + (B_t \cdot Y_t)_G + (G\Sigma_t \cdot Z_t)_{\mathrm{Fr}} + (G\Gamma_t \cdot U_t)_{\lambda_t}\right\} \mathrm{d}t\right].\end{aligned} \tag{2.26}$$

2.3.3 Proof of Theorem 2.1

We now start by proving the uniqueness statement of our main theorem:

Proof of Theorem 2.1 (Uniqueness) Let (X, Y, Z, U, M) and (X', Y', Z', U', M') be two solutions of (2.1)–(2.3). Then, the difference is given by

$$\delta X_t = \int_0^t \delta b_s^\mu \,\mathrm{d}s + \int_0^t \delta\sigma_s^\mu \,\mathrm{d}W_s + \int_{[0,t]\times\mathbf{R}} \delta\gamma_s^\mu(r)\tilde{\eta}^\lambda(\mathrm{d}s, \mathrm{d}r), \tag{2.27}$$

$$\begin{aligned}\delta Y_t =& \delta g_T^\mu + \int_t^T \delta f_s^\mu \,\mathrm{d}s - \int_t^T \delta Z_s \,\mathrm{d}W_s - \int_{(t,T]\times\mathbf{R}} \delta U_s(r)\tilde{\eta}^\lambda(\mathrm{d}s, \mathrm{d}r)\\ &- (\delta M_T - \delta M_t),\end{aligned} \tag{2.28}$$

with $\delta X_0 \equiv 0$. Now, Lemma 2.2 and the definition of $\mathbf{G}^\mu$ in (2.24) yield

$$\mathbb{E}\left[(\delta X_T \cdot \delta g_T^\mu)_G\right] = \mathbb{E}\left[\int_0^T \left(\delta\mathbf{G}_t^\mu \cdot \delta(X_t, Y_t, Z_t, U_t)\right)\mathrm{d}t\right].$$

Then, from the G-mononicity condition (B.6):

$$\begin{aligned}&-\mathbb{E}\left[\int_0^T \left\{\beta_1 \left|G\delta X_t\right|^2 + \beta_2\left(\left|G^\top \delta Y_t\right|^2 + \left\|G^\top \delta Z_t\right\|_{\mathrm{Fr}}^2 + \left\|G^\top \delta U_t\right\|_{\lambda_t}^2\right)\right\}\mathrm{d}t\right]\\ &\quad - \mathbb{E}\left[\beta_3 \left|G\delta X_T\right|^2\right] \geq 0.\end{aligned} \tag{2.29}$$

In the case of $\beta_1, \beta_3 > 0$, from the previous inequality we obtain that $|G\delta X_t|^2 = 0$ $\mathrm{d}\mathbb{P}\otimes\mathrm{d}t$-a.s., which implies that X and X' are $\mathbb{P}$-indistinguishable. In consequence, $Y_T = g_T^\mu(X) = g_T^\mu(X') = Y_T'$. Thus, from the uniqueness of solutions for BSDEs, it follows that $(Y, Z, U, M) = (Y', Z', U', M')$ $\mathbb{P}$-a.s.

In the case of $\beta_2 > 0$, from an analogous argument to the one given in (2.29) we get that $(Y, Z, U) = (Y', Z', U')$ $\mathrm{d}\mathbb{P}\otimes\mathrm{d}t$-a.e.. Thus, by the uniqueness result of (forward) SDEs, $X = X'$ up to $\mathbb{P}$-indistinguishably. In order to verify the equality of $M = M'$, from (2.28) note that

$$\begin{aligned}&\mathbb{E}\left[\left(\delta g_T^\mu + \int_0^T \delta f_t^\mu \,\mathrm{d}t - \delta Y_0 - \int_0^T \delta Z_t \,\mathrm{d}W_t - \int_{(0,T]\times\mathbf{R}} \delta U_t(r)\tilde{\eta}^\lambda(\mathrm{d}t, \mathrm{d}r)\right)^2\right]\\ &= \mathbb{E}\left[\delta M_T^2\right] = \|\delta M\|_{\mathcal{H}^{2,\perp}}^2 = 0.\end{aligned}$$

□

In order to complete the proof of Theorem 2.1 –with the existence part– the following results are needed, providing the ground basis of the continuation argument. Notice that, depending on the relation between the constants $\beta_1, \beta_2, \beta_3$, each case is treated separately; see the comments made in the observations –part c– after Assumptions B.

For the remainder of the proof, we shall use the symbol $\alpha \in [0,1]$ as the index for the parameterized family of coefficients

$$(b,\sigma,\gamma,f)_t^{\mu}(X^\alpha,Y^\alpha,Z^\alpha,U^\alpha) = (b,\sigma,\gamma,f)_t^{\mu,\alpha}.$$

These are coefficients of the system that will be perturbed, adding some elements to the original one depending on the parameter α. On the other hand, the role of the processes B, F, Σ, Γ will become apparent in the iterative procedure proposed after the next results.

Lemma 2.3 *Consider the following two sets of FBSDEs parametrized by $\alpha \in [0,1]$ and $(B, F, \Sigma, \Gamma, \zeta)$ as in (2.25):*

(i) If $\beta_1, \beta_3 > 0$, let:

$$\begin{aligned} X_t^\alpha =& X_0 + \int_0^t \left\{\alpha b_s^{\mu,\alpha} + B_s\right\} \mathrm{d}s + \int_0^t \left\{\alpha\sigma_s^{\mu,\alpha} + \Sigma_s\right\} \mathrm{d}W_s \\ &+ \int_{(0,t]\times\mathbf{R}} \left\{\alpha\gamma_s^{\mu,\alpha}(r) + \Gamma_s(r)\right\} \tilde{\eta}^\lambda(\mathrm{d}s,\mathrm{d}r), \end{aligned} \tag{2.30}$$

$$\begin{aligned} Y_t^\alpha =& Y_T^\alpha + \int_t^T \left\{\alpha f_s^{\mu,\alpha} + F_s + (1-\alpha)\beta_1 G X_s^\alpha\right\} \mathrm{d}s - \int_t^T Z_s^\alpha \,\mathrm{d}W_s \\ &- \int_{(t,T]\times\mathbf{R}} U_s^\alpha(r)\tilde{\eta}^\lambda(\mathrm{d}s,\mathrm{d}r) - (M_T^\alpha - M_t^\alpha), \end{aligned} \tag{2.31}$$

$$X_0 \in L^2(\mathcal{F}_0;\mathbb{R}^d), \quad Y_T = \alpha g^\mu(X_T^\alpha) + \zeta + (1-\alpha)G X_T^\alpha. \tag{2.32}$$

(ii) If $\beta_2 > 0$, let:

$$\begin{aligned} X_t^\alpha =& \int_0^t \left\{\alpha b_s^{\mu,\alpha} + B_s - (1-\alpha)\beta_2\left(G^\top Y_s^\alpha\right)\right\} \mathrm{d}s \\ &+ \int_0^t \left\{\alpha\sigma_s^{\mu,\alpha} + \Sigma_s - (1-\alpha)\beta_2\left(G^\top Z_s^\alpha\right)\right\} \mathrm{d}W_s \\ &+ \int_{(0,t]\times\mathbf{R}} \left\{\alpha\gamma_s^{\mu,\alpha}(r) + \Gamma_s(r) - (1-\alpha)\beta_2\left(G^\top U_s^\alpha(r)\right)\right\} \tilde{\eta}^\lambda(\mathrm{d}s,\mathrm{d}r), \end{aligned} \tag{2.33}$$

$$\begin{aligned} Y_t^\alpha =& Y_T + \int_t^T \left\{\alpha f_s^{\mu,\alpha} + F_s\right\} \mathrm{d}s - \int_t^T Z_s^\alpha \,\mathrm{d}W_s \\ &- \int_{(t,T]\times\mathbf{R}} U_s^\alpha(r)\tilde{\eta}^\lambda(\mathrm{d}s,\mathrm{d}r) - (M_T^\alpha - M_t^\alpha) \end{aligned} \tag{2.34}$$

$$X_0 \in L^2(\mathcal{F}_0; \mathbb{R}^d), \quad Y_T = \alpha g^\mu(X_T^\alpha) + \zeta. \tag{2.35}$$

Assume that there exists $\alpha_0 \in [0,1)$ such that, for each $(B, F, \Sigma, \Gamma, \zeta)$ as in (2.25), there exists a solution $(X^{\alpha_0}, Y^{\alpha_0}, Z^{\alpha_0}, U^{\alpha_0}, M^{\alpha_0})$ of the corresponding FBSDEs in (i) or (ii). Then, there exists a positive constant $\epsilon_0 > 0$ such that, for each $\epsilon \in [0, \epsilon_0]$, with $\epsilon_0 + \alpha_0 \leq 1$, there exists a process $(X^{\alpha_0+\epsilon}, Y^{\alpha_0+\epsilon}, Z^{\alpha_0+\epsilon}, U^{\alpha_0+\epsilon}, M^{\alpha_0+\epsilon})$ solving the associated system of FBSDEs with parameter $\alpha_0 + \epsilon$.

Remark 2.8 Note that by taking $\alpha = 1$ the existence of solutions to (2.30)–(2.32) and (2.33)–(2.35) implies the existence of solutions to (2.1)–(2.3), as we can take $(B, F, \Sigma, \Gamma, \zeta)$ identically zero. Besides, observe that when $\alpha = 0$ the FBSDE reduces to an uncoupled system of the form (2.18)–(2.20) (resp. (2.19)–(2.21)), which has a solution due to Remark 2.7. The next proof will make clear that ϵ_0 does not depend on α_0, which becomes an important observation when the procedure is iterated.

Proof Let α_0 be as in the hypothesis and define the mapping Φ^{α_0} as

$$(B, F, \Sigma, \Gamma, \zeta) \mapsto \Phi^{\alpha_0}(B, F, \Sigma, \Gamma, \zeta) := (X^{\alpha_0}, Y^{\alpha_0}, Z^{\alpha_0}, U^{\alpha_0}, M^{\alpha_0}),$$

where $(X^{\alpha_0}, Y^{\alpha_0}, Z^{\alpha_0}, U^{\alpha_0}, M^{\alpha_0})$ is the solution process to (2.30)–(2.32) associated with the inputs. Now, let

$$(x, y, z, u, m, x_T) \in \mathbb{H}^2(W, \tilde{\eta}^\lambda) \times \mathbb{H}^2(W, \tilde{\eta}^\lambda) \times \mathbb{H}^2(W) \times \mathbb{H}^2(\tilde{\eta}^\lambda) \times \mathcal{H}^{2,\perp} \times L^2(\mathcal{F}_T). \tag{2.36}$$

Using the fact that $(B, F, \Sigma, \Gamma, \zeta)$ has been taken arbitrarily, Remark 2.7 implies that (2.30)–(2.32) has a unique solution process for $(B^\epsilon, F^\epsilon, \Sigma^\epsilon, \Gamma^\epsilon, \zeta^\epsilon)$, defined as

$$\begin{aligned} B_t^\epsilon &:= B_t + \epsilon b_t^\mu(x, y, z, u), & F_t^\epsilon &:= F_t + \epsilon\left(f_t^\mu(x, y, z, u) - \beta_1 G x_t\right), \\ \Sigma_t^\epsilon &:= \Sigma_t + \epsilon \sigma_t^\mu(x, y, z, u), & \zeta^\epsilon &:= \zeta + \epsilon\left(g^\mu(x_T) - G x_T\right), \\ \Gamma_t^\epsilon &:= \Gamma_t + \epsilon \gamma_t^\mu(x, y, z, u), \end{aligned}$$

with $\epsilon > 0$. Composing the previous maps, let $\Phi^{\alpha_0,\epsilon}$ be defined by

$$(x, y, z, u, m, x_T) \mapsto \Phi^{\alpha_0,\epsilon}(x, y, z, u, x_T) : = \Phi^{\alpha_0}(B^\epsilon, F^\epsilon, \Sigma^\epsilon, \Gamma^\epsilon, \zeta^\epsilon). \tag{2.37}$$

Notice that a fix point of $\Phi^{\alpha_0,\epsilon}$ is a solution of the FBSDE system with parameter $\alpha_0 + \epsilon$. Hence, we proceed to prove that, for each $\epsilon \in [0, \epsilon_0]$, this mapping is contractive, with $\epsilon_0 > 0$ to be specified later.

First, let us precise that the domain and range of this mapping is given by

$$\mathcal{H}^2(\mathbb{R}^d) \times \mathcal{H}^2(\mathbb{R}^n) \times \mathbb{H}^2(W) \times \mathbb{H}^2(\tilde{\eta}^\lambda) \times \left(\mathcal{L}^2(W) \oplus \mathcal{L}^2(\tilde{\eta}^\lambda)\right)^\perp .$$

Note that this is a Banach space with respect to a norm equivalent to

$$\|\cdot\|_* = \|\cdot\|_{\mathcal{S}^2(\mathbb{R}^d)} + \|\cdot\|_{\mathcal{S}^2(\mathbb{R}^n)} + \|\cdot\|_{\mathbb{H}^2(W)} + \|\cdot\|_{\mathbb{H}^2(\tilde{\eta}^\lambda)} + \|\cdot\|_{\mathcal{H}^2} ;$$

see Theorems IV.1 and V.2 from [65],

Write

$$\begin{aligned} (X, Y, Z, U, M, X_T) &:= \Phi^{\alpha_0,\epsilon}(x, y, z, u, m, x_T) \quad \text{and} \\ (X', Y', Z', U', M', X'_T) &:= \Phi^{\alpha_0,\epsilon}(x', y', z', u', m', x'_T) \end{aligned}$$

for $(x', y', z', u', m', x'_T)$, another process in the same space as (2.36).

Note that the difference process satisfies

$$\begin{aligned} \delta X_t =& \int_0^t \left\{\alpha_0 \delta b^\mu_s(X, Y, Z, U) + \epsilon \delta b^\mu_s(x, y, z, u)\right\} \mathrm{d}s \\ &+ \int_0^t \left\{\alpha_0 \delta \sigma^\mu_s(X, Y, Z, U) + \epsilon \delta \sigma^\mu_s(x, y, z, u)\right\} \mathrm{d}W_s \\ &+ \int_{(0,t]\times\mathbf{R}} \left\{\alpha_0 \delta \gamma^\mu_s(r, X, Y, Z, U) + \epsilon \delta \gamma^\mu_s(r, x, y, z, u)\right\} \tilde{\eta}^\lambda(\mathrm{d}s, \mathrm{d}r), \end{aligned} \tag{2.38}$$

$$\begin{aligned} \delta Y_t =& \left\{\alpha_0 \delta g^\mu(X_T) + (1-\alpha_0) G \delta X_T + \epsilon \left[\delta g^\mu(x_T) - G \delta x_T\right]\right\} \\ &+ \int_t^T \left\{\alpha_0 \delta f^\mu_s(X, Y, Z, U) + (1-\alpha_0)\beta_1 G \delta X_s\right\} \mathrm{d}s \\ &+ \int_t^T \epsilon \left\{\delta f^\mu_s(x, y, z, u) - \beta_1 G \delta x_s\right\} \mathrm{d}s \\ &- \int_t^T \delta Z_s \,\mathrm{d}W_s - \int_{(t,T]\times\mathbf{R}} \delta U_s(r) \tilde{\eta}^\lambda(\mathrm{d}s, \mathrm{d}r) - (\delta M_T - \delta M_t). \end{aligned} \tag{2.39}$$

Let $q > 0$. From Lemma 2.2 we obtain

$$\begin{aligned} e^{qt}\mathbb{E}\left[(\delta X_t \cdot \delta Y_t)_G\right] =& q \int_0^t e^{qs}\mathbb{E}\left[(\delta X_s \cdot \delta Y_s)_G\right] \mathrm{d}s \\ &- (1-\alpha_0)\beta_1 \int_0^t e^{qs}\mathbb{E}\left[|G\delta X_s|^2\right] \mathrm{d}s \end{aligned}$$

$$
\begin{aligned}
&+\epsilon\beta_1\int_0^t e^{qs}\mathbb{E}\left[(\delta X_s\cdot G\delta x_s)_G\right]\mathrm{d}s\\
&+\alpha_0\int_0^t e^{qs}\mathbb{E}\left[\left(\delta\mathbf{G}_s^{\mu}(X,Y,Z,U)\cdot\delta(X,Y,Z,U)_s\right)\right]\mathrm{d}s\\
&+\epsilon\int_0^t e^{qs}\mathbb{E}\left[\left(\delta\mathbf{G}_s^{\mu}(x,y,z,u)\cdot\delta(X,Y,Z,U)_s\right)\right]\mathrm{d}s.
\end{aligned}
$$

Moreover, from the G-monotonicity and Lipschitz conditions,

$$
\begin{aligned}
&e^{qt}\mathbb{E}\left[(\delta X_t\cdot\delta Y_t)_G\right]\\
&\le q\int_0^t e^{qs}\mathbb{E}\left[(\delta X_s\cdot\delta Y_s)_G\right]\mathrm{d}s-\beta_1\int_0^t e^{qs}\mathbb{E}\left[|G\delta X_s|^2\right]\mathrm{d}s\\
&\quad+\alpha_0\mathbb{E}\left[\int_0^t e^{qs}\left\{-\beta_2\left(\left|G^\top\delta Y_s\right|^2+\left\|G^\top\delta Z_s\right\|_{\mathrm{Fr}}^2+\left\|G^\top\delta U_s\right\|_{\lambda_s}^2\right)\right\}\mathrm{d}s\right]\\
&\quad+\frac{\epsilon}{2}\max\{c_K,\beta_1c_G\}\int_0^t e^{qs}\mathbb{E}\left[|\delta x_s|^2+|\delta y_s|^2+\|\delta z_s\|_{\mathrm{Fr}}^2+\|\delta u_s\|_{\lambda_s}^2\right]\mathrm{d}s\\
&\quad+\frac{\epsilon}{2}\max\{1,\beta_1\}c_G\int_0^t e^{qs}\mathbb{E}\left[|\delta X_s|^2+|\delta Y_s|^2+\|\delta Z_s\|_{\mathrm{Fr}}^2+\|\delta U_s\|_{\lambda_s}^2\right]\mathrm{d}s.
\end{aligned}
$$

Then, Grönwall's inequality yields,

$$
\begin{aligned}
&e^{qt}\mathbb{E}\left[(\delta X_t\cdot\delta Y_t)_G\right]+\beta_1\int_0^t e^{qs}\mathbb{E}\left[|G\delta X_s|^2\right]\mathrm{d}s\\
&\quad+\alpha_0\beta_2\mathbb{E}\left[\int_0^t e^{qs}\left\{\left|G^\top\delta Y_s\right|^2+\left\|G^\top\delta Z_s\right\|_{\mathrm{Fr}}^2+\left\|G^\top\delta U_s\right\|_{\lambda_s}^2\right\}\mathrm{d}s\right]\\
&\le\epsilon\frac{\max\{c_K,c_G,\beta_1c_G\}}{2}e^{qt}\left\{\int_0^t e^{qs}\mathbb{E}\left[|\delta x_s|^2+|\delta y_s|^2+\|\delta z_s\|_{\mathrm{Fr}}^2+\|\delta u_s\|_{\lambda_s}^2\right]\mathrm{d}s\right.\\
&\qquad\left.+\int_0^t e^{qs}\mathbb{E}\left[|\delta X_s|^2+|\delta Y_s|^2+\|\delta Z_s\|_{\mathrm{Fr}}^2+\|\delta U_s\|_{\lambda_s}^2\right]\mathrm{d}s\right\},
\end{aligned}
\tag{2.40}
$$

for all $t\in[0,T]$. On the other hand, due to the terminal conditions,

$$
\begin{aligned}
&\mathbb{E}\left[e^{qT}(\delta X_T\cdot\delta Y_T)_G\right]\\
&\ge[\alpha_0\beta_3+(1-\alpha_0)]\mathbb{E}\left[e^{qT}|G\delta X_T|^2\right]\\
&\quad-\epsilon\mathbb{E}\left[e^{qT}\left\{\frac{|G\delta X_T|^2+|G\delta x_T|^2}{2}-\left(\delta X_T\cdot\delta g^{\mu}(x_T)\right)_G\right\}\right].
\end{aligned}
\tag{2.41}
$$

Note that $\alpha_0\beta_3 + (1 - \alpha_0) > \min\{1, \beta_3\}$. Combining (2.40) and (2.41) and letting $q \to 0$ yields

$$\begin{aligned}
&\overbrace{\min\{1, \beta_1, \beta_3\}}^{>0} \left\{ \|\delta X\|^2 + \|\delta X_T\|_{L^2}^2 \right\} \\
&\le \beta_1 \|\delta X\|^2 + \alpha_0\beta_2 \|\delta(Y, Z, U)\|^2 + [\alpha_0\beta_3 + (1 - \alpha_0)] \|\delta X_T\|_{L^2}^2 \\
&\le \epsilon \frac{\max\{c_K, c_G, \beta_1 c_G\}}{2c_G} \Big\{ \|\delta(x, y, z, u)\|^2 + \|\delta x_T\|_{L^2}^2 \\
&\qquad + \|\delta(X, Y, Z, U)\|^2 + \|\delta X_T\|_{L^2}^2 \Big\}.
\end{aligned} \tag{2.42}$$

where the last scalar product at the r.h.s. of (2.41) is handled by Cauchy's inequality, Young's inequality, and the Lipschitz condition on g:

$$\begin{aligned}
\left(\delta X_T \cdot \delta g^\mu(x_T)\right)_G &\ge - \left|\left(\delta X_T \cdot \delta g^\mu(x_T)\right)_G\right| \\
&\ge - |G\delta X_T| \left|\delta g^\mu(x_T)\right| \ge -\frac{1}{2} \left\{ |G\delta X_T|^2 + K |\delta x_T|^2 \right\}.
\end{aligned}$$

In other words, we have found upper bounds for the forward component of the difference of $\Phi^{\alpha_0,\epsilon}$ in terms of the inputs (x, y, z, u, m) and (x', y', z', u', m').

Next, we proceed to find analogous bounds for the backward component. We do this in two steps: First, due to the BSDE (2.39) and the Lipschitz condition on the terminal cost g, we have that

$$\|\delta Y_T\|^2 \le \underbrace{\underline{\underline{K}} + c_G}_{=:c_{g,G}} \left(\|\delta X_T\|_{L^2}^2 + \epsilon^2 \|\delta x_T\|_{L^2}^2 \right). \tag{2.43}$$

Second, we need an estimate for $\|\delta(Y, Z, U, M)\|^2$. This is done using Lemma 2.1, which describes such type of estimate for BSDEs with different drivers. Indeed: by redefining the drivers of the BSDEs (Y, Z, U, M) and (Y', Z', U', M') as

$$\begin{aligned}
f_s^{1\mu}(Y', Z', U') :=& \alpha_0 f_s^\mu(X, Y', Z', U') + (1 - \alpha_0)GX_s \\
&+ \epsilon(f_s^\mu(x, y, z, u) - \beta_1 Gx_s) + F_s,
\end{aligned}$$

and

$$\begin{aligned}
f^{2\mu}(Y', Z', U') :=& \alpha_0 f_s^\mu(X', Y', Z', U') + (1 - \alpha_0)GX'_s \\
&+ \epsilon(f_s^\mu(x', y', z', u') - \beta_1 Gx'_s) + F_s,
\end{aligned}$$

where we are just modifying the $x-$ component in each one.

By the Lipschitz property of f, these new drivers satisfy the conditions of Lemma 2.1, and hence we obtain

$$
\begin{aligned}
&\|\delta(Y, Z, U, M)\|^2 \\
&\leq c_f \left\{ \|\delta Y_T\|_{L^2}^2 + \mathbb{E}\left[\int_0^T \left| f_t^{1\mu}(Y', Z', U') - f_t^{2\mu}(Y', Z', U') \right|^2 \mathrm{d}t \right] \right\} \\
&\leq c_{f,G} \left\{ \|\delta Y_T\|_{L^2}^2 + \|\delta X\|^2 + \epsilon^2 \|\delta(x, y, z, u, m)\|^2 \right\} \\
&\leq c_{f,g,G} \left\{ \|\delta X_T\|_{L^2}^2 + \|\delta X\|^2 \right\} + c_{f,g,G}\epsilon^2 \left\{ \|\delta x_T\|_{L^2}^2 + \|\delta(x, y, z, u, m)\|^2 \right\},
\end{aligned}
\tag{2.44}
$$

where the last inequality is due to (2.43).

Consequently, by setting

$$
\underline{c} := \frac{(c_{f,g,G} + 1) \max\{c_K, c_G, \beta_1 c_G\}}{2c_G \min\{1, \beta_1, \beta_3\}}
$$

we obtain:

$$
\begin{aligned}
\left\| \delta\Phi^{\alpha_0,\epsilon}(x, y, z, u, m, x_T) \right\|^2 &= \|\delta(X, Y, Z, U, M, X_T)\|^2 \\
&= \|\delta(Y, Z, U, M)\|^2 + \|\delta X\|^2 + \|\delta X_T\|_{L^2}^2 \\
\overset{(2.44)}{\leq} &(c_{f,g,G,\alpha_0} + 1) \left\{ \|\delta X_T\|_{L^2}^2 + \|\delta X\|^2 \right\} + c_{f,g,G}\epsilon^2 \|\delta x_T\|_{L^2}^2 \\
\overset{(2.42)}{\leq} &\underline{c}\epsilon \left\{ \|\delta(x, y, z, u, m, x_T)\|^2 + \|\delta(X, Y, Z, U, M, X_T)\|^2 \right\} \\
&+ c_{f,g,G}\epsilon^2 \|\delta x_T\|_{L^2}^2 .
\end{aligned}
$$

Observe that by taking $\epsilon = \left(2\underline{n}\underline{c}\right)^{-1}$, with $\underline{n} \in \mathbb{N}$, we get $c_{f,g,G}\epsilon^2 \leq \underline{n}^{-2}$. Hence, for $\underline{n} > 1$,

$$
\|\delta(X, Y, Z, U, M, X_T)\|^2 \leq \frac{3}{2\underline{n} - 1} \|\delta(x, y, z, u, m, x_T)\|^2 .
$$

Thus, by taking $\epsilon_0 := \left(6\underline{c}\right)^{-1}$, the mapping $\Phi^{\alpha_0,\epsilon}$ in (2.37) is a contraction for $\epsilon \in [0, \epsilon_0]$.

The proof for the case $\beta_2 > 0$ follows the same procedure, with a slight modification in the definition of the processes $(B^\epsilon, F^\epsilon, \Sigma^\epsilon, \Gamma^\epsilon, \zeta^\epsilon)$, taking instead

$$
B_t^\epsilon := B_t + \epsilon \left(b_t^\mu(x, y, z, u) + \beta_2 G^\top y_t \right), \qquad F_t^\epsilon := F_t + \epsilon f_t^\mu(x, y, z, u),
$$

$$\Sigma_t^\epsilon := \Sigma_t + \epsilon\left(\sigma_t^\mu(x, y, z, u) + \beta_2 G^\top z_t\right), \qquad \zeta^\epsilon := \zeta + \epsilon g^\mu(x_T),$$

$$\Gamma_t^\epsilon := \Gamma_t + \epsilon\left(\gamma_t^\mu(\cdot, x, y, z, u) + \beta_2 G^\top u_t\right).$$

□

Proof of Theorem 2.1 (Existence) We present the proof only for the case $\beta_1, \beta_3 > 0$, since the arguments to prove the other one ($\beta_2 > 0$) are similar and can be adapted by making the corresponding changes. From Remark 2.7 it follows that there exists a solution to (2.30)–(2.32) when $\alpha_0 = 0$ and, by Lemma 2.3 part (i), there exists a constant $\epsilon_0 > 0$ such that, for each $\epsilon \in [0, \epsilon_0]$, there exists a unique solution $(X^\epsilon, Y^\epsilon, Z^\epsilon, U^\epsilon, M^\epsilon)$ of (2.30)–(2.32). Now iterate this procedure k steps, with $\alpha_{i+1} := \alpha_i + \epsilon_0$, and select k in such a way that $k\epsilon_0 \leq 1 < (1 + k)\epsilon_0$, concluding that (2.30)–(2.32) has a unique solution with $\alpha = 1$. Finally, since the initial selection of $(B, F, \Sigma, \Gamma, \zeta)$ was arbitrary, we get the existence of solution to (2.1)–(2.3). □

2.4 Examples and Applications I

In the following, we present a couple of noise processes with a certain intensity that fit our purposes. For the time being, we do not impose a direct relationship between the environment and the dynamics of the system, meaning that we can consider a –possibly– exogenous random environment. Later, we will present some examples to illustrate the applicability of Theorem 2.1 in this case. The particular case of an endogenous environment will be dealt with in the next chapters, when we deal with the McKean-Vlasov equations and mean-field games. We will also return to the examples shown below as a way of presenting a more cohesive view of the most significant results.

2.4.1 Additive Intensity Admissible Noises

Recall first that the probability space we are considering is the one described in (2.22). Given a random environment μ –recall Definition 2.1– satisfying

$$\mathbb{E}\left[\sup_{t\in[0,T]} \mathcal{W}_2(\mu_t, \nu)^2\right] < \infty \tag{2.45}$$

for some (hence, for every) $\nu \in \mathcal{P}_2(\mathbb{R}^d)$, and an $\mathbb{F}^0 \otimes \mathcal{R}$-adapted simple point process η (i.e. $\mathbb{P}[\eta(\{t\} \times C) \in \{0, 1\}] = 1$ for all $t \in [0, T]$ and all $C \in \mathcal{R}$), let

$$\lambda : \Omega^0 \times [0,T] \times \mathcal{R} \times \mathcal{P}_2(\mathbb{R}^d) \times \mathcal{M}_c^*([0,T] \times \mathbf{R}) \to \mathbb{R}_+$$

be a measurable function of the form

$$\begin{aligned}\lambda_t(\omega^0, C, \nu, \eta) &= \psi^0(\nu, C) + \psi^1(\omega^0, t, C)\\ &\quad + \int_{[0,t)\times C} \psi^2(\omega^0, t-s, r)\, \eta(\omega^0, \mathrm{d}s, \mathrm{d}r),\end{aligned} \tag{2.46}$$

where

(i) $\psi^0 : \mathcal{P}_2(\mathbb{R}^d) \times \mathcal{R} \to \mathbb{R}_+$ is a map such that, for each $C \in \mathcal{R}$, $\psi^0(\cdot, C)$ is a non-negative Lipschitz function with respect to the metric $\mathcal{W}_2$ and, for each $\nu \in \mathcal{P}_2(\mathbb{R}^d)$, $\psi^0(\nu, \cdot)$ is a finite measure on $(\mathbf{R}, \mathcal{R})$.
(ii) $\psi^1 : \Omega \times \mathbb{R}_+ \times \mathcal{R} \to \mathbb{R}_+$ is an integrable kernel from $(\Omega^0 \times \mathbb{R}_+, \mathcal{F}^0 \otimes \mathcal{B}(\mathbb{R}_+))$ into $(\mathbf{R}, \mathcal{R})$, independent of μ and η.
(iii) $\psi^2 \in \mathbb{H}^2(\tilde{\eta})$.

Then, from the finiteness of the environment μ specified in (2.45), the condition (3) of Definition 2.2 is satisfied; that is, $K_t^\lambda(\omega^0, C) := \lambda_t(\omega^0, C, \mu_{t-}, \eta^\lambda)$ defines an $\mathbb{F}^0 \vee \mathbb{F}^{\psi^1}$-intensity kernel candidate. Moreover, from the decomposition given by (2.46), we can immediately verify the admissibility of a number of intensity candidates simply by examining each of its terms, as explained below.

1. **Environment-dependant component:** The term ψ^0 allows the environment to directly and explicitly interact with the structure of the jumps, at the cost of requiring a certain level of regularity (Lipschitz) in the space of measures. This type of mappings includes any ψ^0 such that $\psi^0(\cdot, C) = \psi^0(\cdot)$ for all $C \in \mathcal{R}$ (ψ^0 is "independent of the marks") and

$$\psi^0(\nu) = \int_{\mathbb{R}^d} \tilde{\psi}(x)\nu(\mathrm{d}x), \tag{2.47}$$

for some K-Lipschitz function $\tilde{\psi}$. In this case, ψ^0 depends linearly on the measure argument and the Lipschitz property follows directly from an application of Jensen's inequality, as well as the definition of the $\mathcal{W}_2$-distance. Indeed: let $\nu^1, \nu^2 \in \mathcal{P}_2(\mathbb{R}^d)$ and notice that

$$\begin{aligned}\psi^1(\nu^1) &= \int_{\mathbb{R}^d\times\mathbb{R}^d} \tilde{\psi}(x)\, \pi(\mathrm{d}x, \mathrm{d}y) \quad \text{and}\\ \psi^1(\nu^2) &= \int_{\mathbb{R}^d\times\mathbb{R}^d} \tilde{\psi}(y)\, \pi(\mathrm{d}x, \mathrm{d}y)\end{aligned}$$

for all $\pi \in \Pi(\nu^1, \nu^2)$. Then,

$$\left|\psi^1(\nu^1) - \psi^1(\nu^2)\right| = \left|\int_{\mathbb{R}^d \times \mathbb{R}^d} \left\{\tilde{\psi}(x) - \tilde{\psi}(y)\right\} \pi(\mathrm{d}x, \mathrm{d}y)\right|$$
$$\leq \int_{\mathbb{R}^d \times \mathbb{R}^d} \left|\tilde{\psi}(x) - \tilde{\psi}(y)\right| \pi(\mathrm{d}x, \mathrm{d}y)$$
$$\leq K \int_{\mathbb{R}^d \times \mathbb{R}^d} d_{\mathbb{R}^d}(x, y) \pi(\mathrm{d}x, \mathrm{d}y),$$

for all $\pi \in \Pi(\nu^1, \nu^2)$, and therefore, taking the infimum,

$$\left|\psi^1(\nu^1) - \psi^1(\nu^2)\right| \leq K\mathcal{W}_1(\nu^1, \nu^2) \leq K\mathcal{W}_2(\nu^1, \nu^2).$$

With this in mind, two relatively simple examples of valid ψ^0 are

$$m(\nu) := \int_{\mathbb{R}^d} x \, \nu(\mathrm{d}x) \tag{2.48}$$

(i.e. the mean value of ν), and

$$m_k^2(\nu) := \int_{\mathbb{R}^d} \left(k \frac{x}{\max\{k, \|x\|\}}\right)^2 \nu(\mathrm{d}x), \tag{2.49}$$

(i.e. the *truncated* second moment of ν) for some positive constant $k > 0$.

2. **Doubly stochastic component:** Term ψ^1 in (2.46) denotes the *doubly stochastic* or *Cox* component of the kernel. As the name suggests, ψ^1 allows processes with random intensities such as Cox or semi-Cox point processes to be included in the model; see Section 5.2 [20]. For example, let $\lambda = \psi^1$ with ψ^1 assigning measure 1 to jumps of size 1, and zero in any other case (see our previous discussion on counting processes being admissible at the beginning of the chapter). Then, the resulting point process N^λ is a doubly stochastic Poisson with intensity ψ^1.
3. **Self-exciting component:** The term ψ^2 in (2.46) regulates the rates at which the internal history of the process itself affects the appearance of new points/jumps. This means that *some self-exciting (Hawkes-type)* processes can be constructed within our framework [20, Ch. 12].

To better understand how these components interact with each other, we now present an iterative way to obtain an admissible noise process from a candidate that satisfies (2.46). For ease of notation, we consider a one-dimensional, unmarked common noise.

Under the probabilistic structure summarized in (2.22), let N be a Poisson process on $[0, T] \times \mathbb{R}_+$ with intensity measure $\mathrm{d}t \otimes \mathrm{d}s$, and let ψ^i, $i = 0, 1, 2$, satisfy the following:

- $\psi^0 = \lambda_0 \cdot m_{k_0}^2$, with $\lambda_0, k_0 > 0$ positive constants and $m_{k_0}^2$ the function defined in (2.49);

- ψ^1 is a non-negative, bounded, $\mathbb{F}^0$–predictable process on $(\Omega^0, \mathcal{F}^0, \mathbb{P}^0)$, independent from (X_0, μ, N);
- $\psi^2(t) = \lambda \exp\{-kt\}$, with $\lambda, k > 0$ positive constants.

First, by definition of N there exists a sequence of random points in $[0, T] \times \mathbb{R}_+$, denoted by $\{(\tau_n^*, \xi_n^*), n \geq 1\}$ such that $N(\mathrm{d}t, \mathrm{d}s) = \sum_n \delta_{(\tau_n^*, \xi_n^*)}(\mathrm{d}t, \mathrm{d}s)$. We define the first iteration as $\tau_0^{\lambda *} := \tau_0^* \equiv 0$, and

$$\lambda_0^*(\omega^0, t) := \psi^0(\mu_{t-}(\omega^0)) + \psi_t^1(\omega^0) \qquad \forall\, (\omega^0, t) \in \Omega^0 \times (0, T].$$

Then, for each $n \geq 1$ and for all $(\omega^0, t) \in \Omega^0 \times [0, T]$, let

$$\lambda_{n+1}^*(\omega^0, t) := \lambda_n^*(\omega^0, t) + \psi^2(t - \tau_{n+1}^*(\omega^0))\mathbf{1}_{(\tau_{n+1}^*(\omega^0), T]}$$

if $\lambda_n^*(\omega^0, \tau_{n+1}^*(\omega^0)) \geq \xi_{n+1}^*(\omega^0)$, and $\lambda_{n+1}^*(\omega^0, t) := \lambda_n^*(\omega^0, t)$ otherwise.
Lastly, define the intensity candidate λ as the limit

$$\lambda(\omega^0, t) : = \lim_{n \to \infty} \lambda_n^*(\omega^0, t),$$

and the series of jump times $\{\tau_n^\lambda, n \geq 0\}$ as

$$\tau_{n+1}^{\lambda *}(\omega^0) : = \begin{cases} \tau_{n+1}^*(\omega^0), & \text{if } \lambda_n^*(\omega^0, \tau_{n+1}^*(\omega^0)) \geq \xi_{n+1}^*(\omega^0), \\ \tau_n^{\lambda *}(\omega^0), & \text{otherwise;} \end{cases}$$

$$\tau_{n+1}^\lambda(\omega^0) : = \min_{k \geq 0} \left\{\tau_k^{\lambda *}(\omega^0) : \tau_k^{\lambda *}(\omega^0) > \tau_n^\lambda(\omega^0)\right\}.$$

As a result, the process N^λ, given by the jumping times $\{\tau_n^\lambda\}$, admits the $\mathbb{F}^0$–stochastic intensity

$$\lambda_t = \lambda_0 m_{k_0}^2(\mu_{t-}) + \psi_t^1 + \int_{[0,t)} \lambda e^{-k(t-s)}\, \mathrm{d} N_s^\lambda, \tag{2.50}$$

which by construction is dependent on the environment, has an independent doubly stochastic term, and an exponentially decaying self-exciting term.

2.4.2 Environment-Dependent Markov Chains

Another highlight of this methodology is the inclusion of *environment-dependent Markov chains* , a prominent type of dynamical systems closely related to *regime-switching diffusions*, that can be explained in the context of a common Poissonian noise. This type of processes can be regarded as an extension to some Markov renewal processes that were introduced at the beginning of the chapter and which do not necessarily fall into the additive class. The construction is made following [70] and [71].

Let $\mathbf{R} \subset \mathbb{R}^n$ be a finite set of states and, for each $\nu \in \mathcal{P}_2(\mathbb{R}^d)$, let $Q(\nu) = \big(Q(\nu)^{(i,j)}; i, j \in \mathbf{R}\big)$ be the transition semigroup of a continuous-time Markov chain; i.e. Q is a conservative irreducible matrix such that each entry $Q^{(i,j)}$ is a mapping on $\mathcal{P}_2(\mathbb{R}^d)$, which is Lipschitz with respect to the metric $\mathcal{W}_2$. Furthermore, assume that the mapping

$$\mathbf{R} \times \mathcal{P}_2(\mathbb{R}^d) \ni (i, \nu) \mapsto Q(\nu)^{(i)} := \sum_{j \in \mathbf{R}} Q(\nu)^{(i,j)}$$

is uniformly bounded, i.e. there exists a positive constant H_0 such that

$$H_0 := \sup_{\nu \in \mathcal{P}_2(\mathbb{R}^d)} \sup_{i \in \mathbf{R}} Q(\nu)^{(i)} < \infty.$$

Assume that the set $\mathbf{R} \times \mathbf{R}$ is ordered according to the lexicographic ordering [77]. For all $i, j \in \mathbf{R}$ with $i \neq j$ and $\nu \in \mathcal{P}_2(\mathbb{R}^d)$, let $\Gamma^{(i,j)}$ be the consecutive left-closed, right-open intervals of $\mathbb{R}_+$, each having length $Q(\nu)^{(i,j)}$, and $\Gamma^{(i,i)} = \emptyset$:

$$\Gamma(\nu)^{(i,j)} := \left[\underline{Q}(\nu)^{(i,j)}, \underline{Q}(\nu)^{(i,j)} + Q(\nu)^{(i,j)}\right),$$

where

$$\underline{Q}(\nu)^{(i,j)} := \sum_{\mathrm{i} \leq i} \sum_{\substack{\mathrm{j} \leq \underline{j}(\mathrm{i};i) \\ \mathrm{i} \neq \mathrm{j}}} Q(\nu)^{(\mathrm{i},\mathrm{j})}, \qquad \text{with} \qquad \underline{j}(\mathrm{i}; i) = \begin{cases} n, & \mathrm{i} < i, \\ j-1, & \mathrm{i} = i. \end{cases}$$

Given a random environment μ and an independent Poisson process N with intensity measure $\mathrm{d}t \otimes \mathrm{d}r$, let (N^Q, ξ^Q) be an admissible common noise defined (iteratively) by the lifting

$$\eta^Q(\mathrm{d}t, \mathrm{d}r) = \sum_{j \in \mathbf{R}} N\left(\mathrm{d}t, \Gamma(\mu_{t-})^{(i,j)}\right) \otimes \delta_{\{j\}}(\mathrm{d}r),$$

on the event $\{\xi^Q_{N^Q_{t-}} = i\}$, for each $i \in \mathbf{R}$. That is, (N^Q, ξ^Q) admits an intensity kernel given by

$$K^Q(t, \mathrm{d}r) = \sum_{j \in \mathbf{R}} Q(\mu_{t-})^{(i,j)} \, \delta_{\{j\}}(\mathrm{d}r) = Q(\mu_{t-})^{(i,\cdot)} \delta_{\mathcal{E}}(\mathrm{d}r).$$

The process $\xi = \{\xi_t, t \in [0, T]\}$, defined as

$$\xi_t := \xi^Q_{N^Q_t} = \int_{\{t\}\times\mathcal{E}} r\, \eta^Q(\mathrm{d}t, \mathrm{d}r), \tag{2.51}$$

is then referred to as the *environment dependent regime*. That is, ξ is a conditional Markov chain with transition matrix Q:

$$\mathbb{P}\left[\xi_{t+\delta} = j \middle| \xi_t = i, \mathcal{F}^\mu_t\right] = \begin{cases} Q(\mu_t)^{(i,j)}\delta + o(\delta), & i \neq j, \\ 1 + Q(\mu_t)^{(i,i)}\delta + o(\delta), & i = j. \end{cases}$$

With this construction, we can now recover an important class of SDEs commonly known as *Regime Switching Diffusions* [77], i.e. SDEs driven by an idiosyncratic noise process, usually a Brownian motion, where each of the coefficients involved in the dynamic is parametrized by a continuous time, irreducible Markov chain with finite state-space:

Let $b : \mathcal{P}_2(\mathbb{R}^d) \times \mathbb{R}^d \times \mathcal{E} \to \mathbb{R}^d$ be a bounded function such that $(\nu, x) \mapsto b(\nu, x, e)$ satisfy the corresponding *Lipschitz* conditions for all e; let σ be a real constant and $f, g, \gamma \equiv 0$. Then, there exists a strong solution to the forward SDE

$$X^\mu_t = X^\mu_0 + \int_0^t b\left(\mu_s, X^\mu_s, \xi^\mu_s\right) \mathrm{d}s + \sigma W_t,$$

where ξ^μ is the regime determined (as in (2.51)) by the intensity kernel K^Q.

2.4.3 Linear FBSDEs with Self-Exciting Jumps

Each of the previous examples of intensity processes can be useful in recovering well-known models for linear dynamics, extending their dependence on a self-exciting jump process, as in (2.46), with some simplifications. Recall the functions m and $m^2_{k_0}$ introduced in (2.48) and (2.49), respectively.

Let N^λ be an unmarked, self-exciting process with $\mathbb{F}$-stochastic intensity

$$\lambda_t = \lambda_0 m^2_{k_0}(\mu_{t-}) + \psi^1_t + \int_{[0,t)} \psi^2(t-s)\, \mathrm{d}N^\lambda_s, \tag{2.52}$$

with $k_0 > 0$ a truncation parameter, and let $\widehat{b}, \widehat{c}, \widehat{f}, \widehat{g}, \sigma, \widehat{\sigma}, \gamma, \widehat{\gamma} \in \mathbb{R}$ constants such that, for some $\epsilon > 0$, one of the following conditions hold:

$$\widehat{f} - \frac{|\widehat{\sigma}| + |\widehat{\gamma}|}{\epsilon} > 0, \quad \min\{\widehat{c}, (\sigma - \epsilon\,|\widehat{\sigma}|), (\gamma - \epsilon\,|\widehat{\gamma}|)\} \geq 0, \quad g > 0, \tag{2.53}$$

or,

$$\widehat{f} - \frac{|\widehat{\sigma}| + |\widehat{\gamma}|}{\epsilon} \geq 0, \quad \min\{\widehat{c}, (\sigma - \epsilon |\widehat{\sigma}|), (\gamma - \epsilon |\widehat{\gamma}|)\} > 0, \quad g \geq 0. \tag{2.54}$$

We define the coefficients of the forward component as

$$\begin{aligned} b_t^{\mu}(x, y, z, u) &= \widehat{b}(x - m(\mu_{t-})) - \frac{\widehat{c}}{2} y, \\ \sigma_t^{\mu}(x, y, z, u) &= \widehat{\sigma} x - \frac{\sigma}{2} z, \\ \gamma_t^{\mu}(x, y, z, u) &= \widehat{\gamma} x - \frac{\gamma}{2} u, \end{aligned} \tag{2.55}$$

for all $x, y \in \mathbb{R}$. On the other hand, we define the elements of the backward component as

$$f_t^{\mu}(x, y) = \frac{\widehat{f}}{2}(x - m(\mu_{t-})) + \widehat{b} y, \qquad g_T^{\mu}(x) = g\left(x - m(\mu_{T-})\right). \tag{2.56}$$

In other words, we have coupled linear FBSDE

$$\begin{aligned} X_t =& X_0 + \int_0^t \left\{\widehat{b}(X_s - m(\mu_{s-})) - \widehat{c} Y_s\right\} \mathrm{d}s + \int_0^t \left\{\widehat{\sigma} X_s - \frac{\sigma}{2} Z_s\right\} \mathrm{d}W_s \\ &+ \int_{0+}^t \left\{\widehat{\gamma} X_{s-} - \frac{\gamma}{2} U_{s-}\right\} \mathrm{d}\tilde{N}_s^{\lambda}, \end{aligned} \tag{2.57}$$

$$\begin{aligned} Y_t =& g(X_T - m(\mu_{T-})) + \int_t^T \left\{\widehat{f}(X_s - m(\mu_{s-})) + \widehat{b} Y_s\right\} \mathrm{d}s - \int_t^T Z_s \,\mathrm{d}W_s \\ &- \int_{t+}^T U_{s-} \,\mathrm{d}\tilde{N}_s^{\lambda} - (M_T - M_t). \end{aligned} \tag{2.58}$$

It can be verified easily that the coefficients in (2.55) and (2.56) satisfy the conditions from Theorem 2.1. Indeed, let $G = 1$ and $\mathbf{G}^{\mu}$ be given by

$$\mathbf{G}_t^{\mu}(x, y, z, u) := \begin{bmatrix} -\frac{\widehat{f}}{2}(x - m(\mu_t)) - \widehat{b} y \\ \widehat{b}(x - m(\mu_t)) - \frac{\widehat{c}}{2} y \\ \widehat{\sigma} x - \frac{\sigma}{2} z \\ \widehat{\gamma} x - \frac{\gamma}{2} u \end{bmatrix}.$$

Then, direct calculations yield

$$\begin{aligned} &\left(\delta \mathbf{G}_t^{\mu}(x, y, z, u) \cdot \delta(x, y, z, u)\right) \\ &= -\frac{\widehat{f}}{2} |\delta x|^2 - \frac{\widehat{c}}{2} |\delta y|^2 + \widehat{\sigma}(\delta x \cdot \delta z) - \frac{\sigma}{2} |\delta z|^2 + \widehat{\gamma}(\delta x \cdot \delta u)_{\lambda_t} - \frac{\gamma}{2} \|\delta u\|_{\lambda_t}^2 \end{aligned}$$

$$
\begin{aligned}
&\leq -\frac{1}{2}\left(\widehat{f} - \frac{|\widehat{\sigma}| + |\widehat{\gamma}|}{\epsilon}\right)|\delta x|^2 \\
&\qquad - \frac{1}{2}\widehat{c}\,|\delta y|^2 - \frac{1}{2}\left(\sigma - \epsilon\,|\widehat{\sigma}|\right)|\delta z|^2 - \frac{1}{2}\left(\gamma - \epsilon\,|\widehat{\gamma}|\right)\|\delta u\|^2_{\lambda_t},
\end{aligned}
$$

where the last inequality due to Young's inequality; thus, by taking

$$
\beta_1 = \widehat{f} - \frac{|\widehat{\sigma}| + |\widehat{\gamma}|}{\epsilon}, \qquad \beta_2 = \min\left\{\widehat{c}, \left(\sigma - \epsilon\,|\widehat{\sigma}|\right), \left(\gamma - \epsilon\,|\widehat{\gamma}|\right)\right\} \qquad \beta_3 = g,
$$

as constants in the G–monotonicity condition, along with either (2.53) or (2.54), we conclude that there exists a unique process

$$
(X, Y, Z, U, M) \in \mathcal{H}^2(\mathbb{R}) \times \mathcal{H}^2(\mathbb{R}) \times \mathbb{H}^2(W) \times \mathbb{H}^2(\tilde{\eta}^\lambda) \times \mathcal{H}^{2,\perp}(\mathbb{R})
$$

satisfying (2.57) and (2.58), subject to the fixed random environment μ with stochastic intensity λ given by (2.52).

Chapter 3
Optimization in Random Environments

Abstract Controlled processes are the main tool of interest in this chapter, where the goal consists of minimizing a cost functional depending on the trajectories of the state process (a forward SDE) over the set of admissible control processes. The approach to solving this, characterizing an optimal control and establishing the necessary and sufficient conditions for their existence, is to use the maximum principle. The coupled system of FBSDEs characterizing the minimizing control are solved using the results from Chap. 2. Some illustrative examples are also included to highlight the application of this methodology.

3.1 Problem Formulation

In this chapter, we focus on solving the following optimal control problem for an arbitrary (but *fixed*) random environment μ: Given a nonempty Borel set A and an admissible set-up $(X_0, \mu, \lambda, (N^\lambda, \xi), W)$, we consider the problem of

$$\underset{\alpha\in\mathcal{A}}{\text{minimize}}\ J^\mu(X,\alpha), \quad \text{with} \quad J^\mu(X,\alpha) := \mathbb{E}\left[\int_0^T f^\mu(X^\mu_{t-},\alpha_{t-})\,\mathrm{d}t + g^\mu(X^\mu_{T-})\right], \tag{3.1}$$

where $\mathcal{A}$ denotes *set of admissible strategies*, to be specified below, and X^μ is subject to the dynamics.

$$\begin{aligned} X^\mu_t =& X_0 + \int_0^t b^\mu\left(X^\mu_{s-},\alpha_{s-}\right)\mathrm{d}s + \int_0^t \sigma^\mu\left(X^\mu_{s-},\alpha_{s-}\right)\mathrm{d}W_s \\ &+ \int_{(0,t]\times\mathbf{R}} \gamma^\mu\left(r, X^\mu_{s-},\alpha_{s-}\right)\tilde{\eta}^\lambda(\mathrm{d}s,\mathrm{d}r). \end{aligned} \tag{3.2}$$

D. Hernández-Hernández, J. H. Ricalde-Guerrero, *Optimal Control in Random Environments*, SpringerBriefs in Mathematics,
https://doi.org/10.1007/978-3-032-25641-6_3

Each pair (X^α, α) is then referred to as an α–controlled pair. If, in addition, $(X^{\widehat{\alpha}}, \widehat{\alpha}) \equiv (\widehat{X}, \widehat{\alpha})$ is a minimizer of J^μ in (3.1), we refer to it as the *optimal pair* of the problem.

Remark 3.1 The choice of terminal cost $\mathbb{E}[g(\mu_{T-}, X_{T-})]$ is mainly motivated by practicality, as it produces a predictable terminal condition. However, due to the Poissonian nature of the jumps, this choice is indistinguishable from $\mathbb{E}[g(\mu_{T-}, X_T)]$, since there is no fixed discontinuity and the set of jump times is *totally inaccessible* (recall the elementary properties of Poisson point processes presented in Sect. 1.3). As a consequence, we have $\mathbb{P}[X_t \neq X_{t-}] = 0$ for every $t \in (0, T]$.

Our approach is based on the so-called *Pontryagin's Maximum Principle* (PMP for short). Namely, we use variational arguments to develop a set of necessary conditions that an optimal trajectory must fulfill, as well as an adjoint set of differential equations which capture the variation of the payoff functional $J^\mu(X, \alpha)$ when the control α is slightly perturbed. Technically, our method is motivated by the works of Tang and Li [73] and Framstad et al. [35], which considered systems driven by jump diffusion, but in a simpler setting, that is, with no random environment. Lastly, to ease our notation, we write

$$\widehat{b}_t^\mu := b(\mu_{t-}, \widehat{X}_{t-}, \widehat{\alpha}_{t-}), \qquad \mathrm{d}\mathbb{P} \otimes \mathrm{d}t - \text{a.e.} \tag{3.3}$$

(and similarly for $\sigma, \gamma, \lambda, \ldots$) when evaluating in the optimal $\widehat{\alpha}$-controlled pair, with the differences being represented by

$$\delta\widehat{b}_t^\mu(x, a) = b(\mu_{t-}, x, a) - b(\mu_{t-}, \widehat{X}_{t-}, \widehat{\alpha}_{t-}), \qquad \mathrm{d}\mathbb{P} \otimes \mathrm{d}t - \text{a.e.},$$

for all $(x, a) \in \mathbb{R}^d \times A$.

3.1.1 The Stochastic Case

Continuing from the brief reminder of the deterministic case done in Sect. 1.2, we now turn to the study of stochastic optimal control problems, examining the stochastic version of the PMP; these formulations can be found in classical works by Bismut [14], Bensoussan [12], Peng [61], El-Karoui et al. [34], to name a few. As in the deterministic case, most of the elements are retained here: the presence of an adjoint backwards system (the costate Y), a Hamiltonian H, and the Pontryagin's condition. However, as these elements are now randomized, some modifications are appropriate.

Remark 3.2 In order to maintain the framework presented in the previous chapter, we focus only on the *strong formulation* of the problem. That is, we keep the reference probability space fixed throughout, so that a solution to an FBSDE is a process defined on the same space; recall Definition 2.2.

Let $(\Omega, \mathcal{F}, \mathbb{P})$ be a probability space, let W be a standard Brownian Motion defined on the same space, and let $\mathbb{F} \equiv \mathbb{F}^W$ be the (complete, right continuous) filtration generated by W. Instead of defining a control as in (1.1), we say that α is an *admissible control* if

$$\alpha : [0, T] \times \Omega \to A \text{ is adapted to } \mathbb{F}.$$

In the context of stochastic processes, this means that a control is admissible if we can determine its value at any moment $t \in [0, T]$ simply by keeping track of the information provided up to that point by the system itself (represented by the filtration $\mathbb{F}$).

Similarly, the dynamic of the *controlled state* (1.2) is replaced by a W-driven SDE; i.e., X^α is now defined as a stochastic process in $(\Omega, \mathcal{F}, \mathbb{P})$:

$$\begin{aligned} \mathrm{d}X_t^\alpha &= b(X_t^\alpha, \alpha_t)\,\mathrm{d}t + \sigma(X_t^\alpha)\,\mathrm{d}W_t, \quad 0 \le t \le T, \\ X_0 &\in L^2(\mathbb{R}^d), \end{aligned} \tag{3.4}$$

for two suitable functions b and σ, and $\mathrm{d}W_t$ represents an integral in the sense of Itô.

In this case, the controlled system does not correspond to a single trajectory in the phase space but to a family of trajectories indexed by $\omega \in \Omega$ and "weighted" by $\mathbb{P}$ (without lost of generality, we can take $\mathbb{P}$ as the Wiener measure over the set of continuous functions, thus rendering the problem as one of optimization in infinite dimensional spaces). As a result, the notion of finding the control for "the best strategy among all the possible paths" becomes inadequate. Instead, the objective now becomes finding the control that minimizes the objective "on average":

$$\widehat{\alpha} \in \arg\min \mathbb{E}\left[\int_0^T f(X_t^\alpha, \alpha_t)\,\mathrm{d}t + g(X_T^\alpha)\right], \tag{3.5}$$

where $\mathbb{E}$ denotes the expectation w.r.t. $\mathbb{P}$. To ease the notation, we use the same symbol $J(\alpha)$ to represent the objective function in (3.5), with the understanding that we are in the stochastic case.

Following the steps posed by its original formulation, the Stochastic Maximum Principle (SMP) seeks to establish a system of differential equations that an optimal control $\widehat{\alpha}$, in the sense of (3.5), must satisfy. In order to achieve this aim, observe that the controlled state X is itself random. Then, by defining the variation of the state with respect to the control as the corresponding variation on a given realization of the process (i.e. indexed by the events $\omega \in \Omega$), we can deduce that the resulting costate process Y is also random.

Moreover, recall that in the deterministic case, the optimal control $\widehat{\alpha}$ necessarily satisfies Pontryagin's Condition (1.8), which itself is dependent of both the state X

and the costate Y through the Hamiltonian. This was enough in the deterministic case because no additional measurability condition was imposed on the control, whereas in the stochastic case the adaptability of Y to $\mathbb{F}$ must be taken into account (the adaptability of X is already guaranteed due to the Itô integral term in (3.4)). In other words, to ensure the $\mathbb{F}$-adaptability of any optimal control candidate, Y itself must be $\mathbb{F}$-adapted. Consequently, for Y to maintain its role as a conjugate process of X, the SMP must consider the notion of orthogonality between stochastic processes, which in turn leads back to the notion of martingale decomposition and the need to consider backward SDEs.

Recall that for the moment we are assuming $\mathbb{F} = \{\mathcal{F}_t, 0 \leq t \leq T\}$ to be generated only by W. Then, from Chap. 2 (see Proposition 2.1), we know that for any $\mathcal{F}_T$-measurable random variable ζ with finite second moment there exists an $\mathbb{F}$-adapted process Z such that

$$\zeta = \mathbb{E}\left[\zeta | \mathcal{F}_0\right] + \int_0^T Z_t \,\mathrm{d}\, W_t.$$

Returning to the role of Y as a costate, we also know that, for the separation plane to be "pointing in the right direction", Y must also satisfy the terminal condition $Y_T = \mathrm{D}_x\, g(X_T^{\widehat{\alpha}})$. This means that, unlike the deterministic case, there is an additional martingale term $Z_t \,\mathrm{d}\, W_t$ present in the dynamics of Y. Thus, by checking the duality relation between X and Y as L^2 processes (that is, w.r.t. to the L^2 inner product), we can interpret Z as the conjugate of the "derivative of X in the direction of W", given by $\sigma(X_t^{\widehat{\alpha}})$ in (3.4). In fact, this idea can be formalized by the notion of derivatives *with respect to a Brownian motion* provided by the theory of Malliavin calculus; see, for example, [2].

Following this set of ideas, we see that the supporting hyperplane satisfies the BSDE

$$\begin{aligned}
\mathrm{d}\, Y_t &= -\left\{(Y_t \cdot \mathrm{D}_x\, b(X_t^{\widehat{\alpha}}, \widehat{\alpha}_t)) + (Z_t \cdot \mathrm{D}_x\, \sigma(X_t^{\widehat{\alpha}})) + \mathrm{D}_x\, f(X_t^{\widehat{\alpha}}, \widehat{\alpha}_t)\right\} \mathrm{d}\, t + Z_t \,\mathrm{d}\, W_t, \\
Y_T &= \mathrm{D}_x\, g(X_T^{\widehat{\alpha}})
\end{aligned}$$

for all $0 \leq t \leq T$ $\mathbb{P}$—a.e. in Ω. Consequently, by defining the Hamiltonian H as

$$H(x, y, z, a) := (y \cdot b(x, a)) + (z \cdot \sigma(x)) + f(x, a),$$

the SMP states that if $\left(X^{\widehat{\alpha}}, \widehat{\alpha}\right)$ is an optimal pair, then there exists a pair of processes (Y, Z) such that $(X^{\widehat{\alpha}}, Y, Z)$ satisfies the FBSDE

$$\mathrm{d}\, X_t^{\widehat{\alpha}} = \mathrm{D}_y\, H(X_t^{\widehat{\alpha}}, Y_t, Z_t, \widehat{\alpha}_t)\, \mathrm{d}\, t + \mathrm{D}_z\, H(X_t^{\widehat{\alpha}}, Y_t, Z_t, \widehat{\alpha}_t)\, \mathrm{d}\, W_t, \tag{3.6}$$

$$\mathrm{d}\, Y_t = -\,\mathrm{D}_x\, H(X_t^{\widehat{\alpha}}, Y_t, \widehat{\alpha}_t)\, \mathrm{d}\, t + Z_t \,\mathrm{d}\, W_t, \tag{3.7}$$

$$X_0^{\widehat{\alpha}} \in L^2(\mathbb{R}^d), \qquad Y_T = \mathrm{D}_x\, g(X_T^{\widehat{\alpha}}). \tag{3.8}$$

for every $t \in [0, T]$, $\mathbb{P}$–a.e. in Ω, and $\widehat{\alpha}$ satisfies the Pontryagin condition

$$H(X_t^{\widehat{\alpha}}, Y_t, Z_t, \widehat{\alpha}_t) \leq H(X_t^{\widehat{\alpha}}, Y_t, Z_t, a), \qquad \forall a \in A, \tag{3.9}$$

for every $t \in [0, T]$, $\mathbb{P}$—a.e. in Ω.

It is worth noting that the PMP (or SMP) is not the only method to tackle optimization problems. Bellman's Dynamic Programming Principle [9–11] is another widely known and useful approach, which dates back to the same era as the original PMP, sharing similar motivations. For a comparison between both methods, we refer to the book by Yong and Zhu [78], and references therein. However, the tools it provides have made it one of the most widely adopted methods to solve optimal control problems. This, in turn, has led to many extensions, including an SMP for discontinuous processes [35, 73]. For the sake of brevity, we omit the results presented there, as a generalization to include random environments will also be presented later in the chapter, but we do want to highlight the work done by these authors.

3.1.2 The Linear-Quadratic Model: A Motivating Example

Before proceeding to the main setting, we present a relatively simple case based on the system introduced in Sect. 2.4.3 of the previous chapter, in order to exemplify how the SMP can be used to tackle optimization problems in the form of (3.1)–(3.2). Specifically, we assume that the dynamics of the state are given by a linear SDE, and the costs in the payoff are quadratic. In the literature, this is the well-known *linear-quadratic* (LQ) optimal control problem. We choose this particular example as the LQ model is a well-known benchmark in the theory of optimal control since it illustrates some of the benefits and complexities of the method, and in some cases it allows to have explicit or semiexplicit computations.

Let A be a nonempty convex set, and let $(\Omega, \mathcal{F}, \mathbb{F}, \mathbb{P})$ be a probability space with an admissible set-up $(X_0, \mu, \lambda, (N^\lambda, \xi), W)$, where μ is an exogenous random environment. For illustrative purposes, we are assuming that the processes involved are 1-dimensional, with $A \subseteq \mathbb{R}$. We also assume $(N^\lambda, \xi) \equiv N^\lambda$ is an unmarked Poissonian common noise with intensity λ such that $\lambda_0 = k_0$ and

$$\lambda_t = m_{k_0}^2(\mu_{t-}) + k_1 + \int_{[0,t)} e^{-k_2(t-s)}\, \mathrm{d} N_s^\lambda, \qquad \forall t > 0, \tag{3.10}$$

where $k_0, k_1, k_2 > 0$ are given parameters, and $m_k^2(\nu)$ denotes the truncated second moment of ν, as defined in (2.49). That is, λ is an additive intensity, dependent on the truncated second moment of the environment and with an exponentially decaying self-exciting component; see Sect. 2.4.1.

Given real constants $b, \sigma, \gamma, f, f_1, f_2, g$, consider the optimal control problem with random environment μ defined by minimization of the payoff

$$J^{\mu}(X,\alpha)=\mathbb{E}\Big[\int_0^T\Big\{\frac{f_1}{2}(X_{t-}-m(\mu_{t-}))^2+f(X_{t-}-m(\mu_{t-}))\alpha_t+\frac{f_2}{2}\alpha_t^2\Big\}\mathrm{d}t + \frac{g}{2}(X_{T-}-m(\mu_{T-}))^2\Big], \tag{3.11}$$

where m is the mean function of (2.48), and the state X evolves according to the controlled dynamics

$$X_t = X_0+\int_0^t\{b(X_{s-}-m(\mu_{s-}))+\alpha_s\}\mathrm{d}s+\sigma W_t+\gamma\tilde{N}_t^{\lambda}, \tag{3.12}$$

for any $\mathbb{F}$-predictable process $\alpha:\Omega\times[0,T]\to A$.

In the spirit of the SMP, the objective now would be to deduce the Hamiltonian of the problem to find the corresponding candidates for optimal control. However, the difficulty here comes from the fact that the intensity of the common noise N^{λ} depends on the environment λ. This means that the conjugate process of γ lies in the space of integrands $\mathfrak{H}^{\lambda,\mu}$ defined in (2.9), and therefore now we must consider the random inner product $(\,\cdot\,,\,\cdot\,)_{\lambda}$ of (2.10) within the Hamiltonian:

$$\begin{aligned}H_t^{\mu}(x,y,z,u,a):=&\frac{f_1}{2}(x-m(\nu))^2+f(x-m(\nu))a+\frac{f_2}{2}a^2\\&+y\cdot\{b(x-m(\nu))+a\}+z\cdot\sigma+\underbrace{u\cdot\lambda_t\cdot\gamma}_{=:(u\cdot\gamma)_{\lambda_t}},\end{aligned} \tag{3.13}$$

on the event $\{\mu_{t-}=\nu\}$.

Then, extending the intuition provided by the usual SMP, we would expect to find that the Linear-Quadratic Optimal Control problem given by (3.11)–(3.10) is solvable if, and only if, there exists a solution to the FBSDE

$$\widehat{X}_t = X_0+\int_0^t\big\{b\big(\widehat{X}_s-m(\mu_{s-})\big)+\widehat{\alpha}_s\big\}\mathrm{d}s+\sigma W_t+\gamma\tilde{N}_t^{\lambda}, \tag{3.14}$$

$$\begin{aligned}\widehat{Y}_t =& g\big(\widehat{X}_T-m(\mu_{T-})\big)+\int_t^T\big\{f_1\big(\widehat{X}_s-m(\mu_{s-})\big)+f\widehat{\alpha}_s+b\widehat{Y}_t\big\}\mathrm{d}s\\&-\int_t^T\widehat{Z}_s\,\mathrm{d}W_s-\int_t^T\widehat{U}_s\,\mathrm{d}\tilde{N}_s^{\lambda},\end{aligned} \tag{3.15}$$

where $\widehat{\alpha}$ is the admissible control given by inequality

$$H_t^{\mu}\big(\widehat{X},\widehat{Y},\widehat{Z},\widehat{U},\widehat{\alpha}\big)\le H_t^{\mu}\big(\widehat{X},\widehat{Y},\widehat{Z},\widehat{U},\alpha\big) \tag{3.16}$$

for every $\mathbb{F}$-predictable process $\alpha:\Omega\times[0,T]\to A$.

In solving the posed stochastic optimal control problem, observe that (3.10) and (3.14)–(3.15) form a linear system of FBSDEs similar to the one studied in

Sect. 2.4.3. Thus, if (3.14)–(3.15) are in fact a *Hamiltonian system* akin to the one in the deterministic or stochastic cases—Eqs. (1.6)–(1.7) or (3.6)–(3.8) respectively—, we need only to verify the G-monotonicity condition from Theorem 2.1 of the previous chapter in order to identify the optimal trajectories of the state and costate processes. However, this still leaves us with the problem of determining wether or not (3.16) determines the optimality of the control, i.e. if it truly corresponds to a *maximum principle*.

The main results of this chapter—Theorems 3.1 and 3.2 below—consist in formalizing the intuition behind the SMP to the setting where an environment induces discontinuities on the system. In this way, optimal control problems such as the one presented above can be solved with techniques similar to the ones already existent in literature. Regarding the LQ model itself, we will come back to it later in Sect. 4.6, once all of our results are established, and characterize its solutions along with its mean-field game counterpart.

3.2 Hamiltonian System: Definitions and Assumptions

The method presented next to verify the SMP is based on [73], and systematically uses variational arguments to obtain the Hamiltonian system corresponding to the game problem in a random environment. This means that, given an optimal pair $(\widehat{X}, \widehat{\alpha})$ for the optimization problem in (3.1)–(3.2), our goal is to derive the set of adjoint BSDEs which captures the first and second order variations of the optimally controlled system $\widehat{X}$, given a small perturbation of the optimal control $\widehat{\alpha}$. The Hamiltonian, defined below, is a crucial tool for this purpose, in order to describe the variations in the total cost J^{μ} when a representative player deviates from the optimally controlled trajectory $\widehat{X}$.

Let μ be an arbitrary random environment. The *Hamiltonian* is defined as the measurable mapping

$$H : \Omega \times [0, T] \times \mathcal{P}_2(\mathbb{R}^d) \times \mathbb{R}^d \times \mathbb{R}^d \times \mathbb{R}^{d\times k} \times \mathfrak{H}^{\lambda,\mu} \times A \to \mathbb{R}$$

given by

$$\begin{aligned} &H(\omega, t, \nu, x, y, z, U_t(\omega, \cdot), a) \\ &\quad := f(\nu, x, a) + (y \cdot b(\nu, x, a)) + (z \cdot \sigma(\nu, x, a))_{\mathrm{Fr}} \\ &\qquad + \int_{\mathbf{R}} \mathrm{Tr}(U_t(\omega, r)\,\mathrm{diag}\,(\lambda(r, \mu_{t-}(\omega))Q(\mathrm{d}r))\,\gamma(r, \nu, x, a)^{\top}). \end{aligned}$$

We also define the Jacobian $\mathrm{D}_x H$ and the Hessian $\mathrm{D}_x^2 H$ accordingly.

Remark 3.3 Implicit in the definition of Hamiltonian there is the assumption that we are considering the case $d = d_0$ from the previous chapter. We also omit the subindex x when writing the Jacobian and the Hessian—$\mathrm{D}\,H \equiv \mathrm{D}_x\,H$ and $\mathrm{D}^2\,H \equiv \mathrm{D}_x^2\,H$, respectively—, with the implicit convention that we shall be writing the results of the chapter in terms of differentials on the x axis.

Then, the *path-wise admissible Hamiltonian* refers to the predictable random field $H^\mu = \{H_t^\mu, 0 \leq t \leq T\}$ where

$$(\omega, t) \mapsto \underbrace{H(\omega, t, \mu_{t-}(\omega), x, y, z, U_t(\omega, \cdot), a)\mathbf{1}_{\{U_t(\omega,\cdot)=u(\cdot)\}}(\omega),}_{=: \, H_t^\mu(x, y, z, u, a)(\omega)} \tag{3.17}$$

is $\mathbb{F}$-predictable for any $(x, y, z, a) \in \mathbb{R}^d \times \mathbb{R}^d \times \mathbb{R}^{d\times k} \times A$ and any $u : \mathbf{R} \to \mathbb{R}^{d\times n_0}$ with $\|u(\cdot)\|_{\lambda_t^\mu(\omega)} < \infty$. Picking up from the previous notation, we write

$$H_t^\mu(x, y, z, u, a) = H_t^\nu(x, y, z, u, a) \text{ on the event } \{\mu_{t-} = \nu\} \text{ for any } \nu \in \mathcal{P}_2(\mathbb{R}^d);$$

see (2.9). The processes $\{\mathrm{D}\,H_t^\mu, 0 \leq t \leq T\}$ and $\{\mathrm{D}^2\,H_t^\mu, 0 \leq t \leq T\}$ are defined in a similar way and will be stated explicitly later. It is also worth noting that due to the characterization of the jump process via its compensator—or equivalently its intensity—, the jump components interact with each other only through the inner product induced by predictable quadratic covariation of η^λ, i.e. $(\cdot, \cdot)_{\lambda_t^\mu}$ from (2.10), thus making H a function that does not depend on $r \in \mathbf{R}$ directly.

Other relevant assumptions that will be consider in this chapter are as follow: Conditions (C.1)–(C.2) are technical conditions on the setting that ensures its admissibility, thus allowing us to make use of the results from the previous chapter. Condition (C.3) is an additional condition that we introduce here and is related to the existence of the required a priori estimates—compare with (2.45) and the notion of additive intensities from Sect. 2.4.1. Conditions (C.4)–(C.7) ensure that the Hamiltonian system, described below in terms of the adjoint FBSDE, is well-posed [38, 39]; namely, they correspond to the controlled version of the coefficients from the previous chapter, with the *monotonicity condition* (C.7) in particular based upon the work of Wu [79] for the existence and uniqueness of strong solutions for FBSDEs with jumps. Lastly, (C.8)–(C.9) are technical conditions used to ensure Pontryagin's (maximal) condition is satisfied by the optimal control. Recall that we are using the δx notation to denote the difference $x - x'$.

Assumptions C
(i) On the probability space

(C.1) *The tuple $(X_0, \mu, \lambda, (N^\lambda, \xi), W)$ is a given admissible set-up—in the sense of Definition 2.2 from Chap. 2—for some Polish space $(\Omega, \mathcal{F})$ endowed with its Borel σ-algebra.*

(C.2) *The intensity candidate has the form $\lambda(r, \mu_{t-})$ for some measurable deterministic function $\lambda : \mathbf{R} \times \mathcal{P}_2(\mathbb{R}^d) \to \mathbb{R}_+^{n_0}$. Furthermore, λ is a Lipschitz function on $\mathcal{P}_2(\mathbb{R}^d)$ with respect to the 2-Wasserstein distance, uniformly continuous on r.*

(C.3) The random environment μ satisfies the uniform bound

$$\|\mu\|_{\mathcal{W}} := \sup_{[0,T]} \left(\mathbb{E}\left[\mathcal{W}_2\left(\mu_t, \delta_{\{0\}}\right)^8\right]\right)^{\frac{1}{8}} < \infty. \tag{3.18}$$

(ii) On the coefficients

(C.4) ***Second differentiability on the state***

- $(b, \sigma, f, g)(\nu, \cdot, a) \in \mathrm{C}_b^2(\mathbb{R}^d)$ *for all* $(\nu, a) \in \mathcal{P}_2(\mathbb{R}^d) \times A$;
- $\gamma(r, \nu, \cdot, a) \in \mathrm{C}_b^2(\mathbb{R}^d)$ *for all* $(r, \nu, a) \in \mathbf{R} \times \mathcal{P}_2(\mathbb{R}^d) \times A$;
- $g(\nu, \cdot) \in \mathrm{C}_b^2(\mathbb{R}^d)$ *for all* $\nu \in \mathcal{P}_2(\mathbb{R}^d)$.

(C.5) ***Uniform continuity on the control***

- $(\Psi, \mathrm{D}\,\Psi, \mathrm{D}^2\,\Psi)(\nu, x, \cdot) \in \mathrm{UC}(A)$ *for all* $(\nu, x) \in \mathcal{P}_2(\mathbb{R}^d) \times \mathbb{R}^d$, *where* Ψ *can be either* b, σ *or* f;
- $(\gamma, \mathrm{D}\,\gamma, \mathrm{D}^2\,\gamma)(r, \nu, x, \cdot) \in \mathrm{UC}(A)$ *for all* $(r, \nu, x) \in \mathbf{R} \times \mathcal{P}_2(\mathbb{R}^d) \times \mathbb{R}^d$.

(C.6) ***Lipschitz condition on the measure and the adjoint states*** *Let μ' be another random environment such that $\mathfrak{H}^{\lambda,\mu} \cap \mathfrak{H}^{\lambda,\mu'} \neq \emptyset$. There exists a positive constant $K > 0$, such that for $\mathrm{d}\,t$-almost all $t \in [0, T]$ and all $a \in A$, the following inequality holds for $\mathbb{P}$-almost surely:*

$$\begin{aligned}
&\left|b_t^{\nu}(x,a) - b_t^{\nu'}(x',a)\right|^2 + \left\|\sigma_t^{\nu}(x,a) - \sigma_t^{\nu'}(x',a)\right\|_{\mathrm{Fr}}^2 \\
&\quad + \left\|\gamma_t^{\nu}(x,a) - \gamma_t^{\nu'}(x',a)\right\|_{\lambda_t^{\mu}}^2 \\
&\quad + \left|\mathrm{D}\,H_t^{\nu}(x,y,z,u,a) - \mathrm{D}\,H_t^{\nu'}(x',y',z',u',a)\right|^2 \\
&\quad + \left|\mathrm{D}\,g^{\nu}(x) - \mathrm{D}\,g^{\nu'}(x)\right|^2 \\
&\leq K\Big\{\mathcal{W}_2(\nu,\nu')^2 + \left|x - x'\right|^2 + \left|y - y'\right|^2 \\
&\quad + \left\|z - z'\right\|_{\mathrm{Fr}}^2 + \left\|u - u'\right\|_{\lambda_t^{\mu}}^2\Big\},
\end{aligned}$$

with H as in (3.17), *for every*

$$\begin{aligned}
&(\nu, x, y, z, U), (\nu', x', y', z', U') \\
&\quad \in \mathcal{P}_2(\mathbb{R}^d) \times \mathbb{R}^d \times \mathbb{R}^d \times \mathbb{R}^{d\times m} \times (\mathfrak{H}^{\lambda,\mu} \cap \mathfrak{H}^{\lambda,\mu'})
\end{aligned}$$

on the event $\{\mu_{t-} = \nu,\ U_t = u\} \cap \{\mu'_{t-} = \nu',\ U'_t = u'\}$.

(C.7) ***Monotonicity condition on the adjoint states*** *There exist nonnegative constants* $\beta_1, \beta_2, \beta_3$ *with* $\beta_1 + \beta_2 > 0$ *and* $\beta_2 + \beta_3 > 0$ *such that for* $\mathrm{d}t$*-almost all* $t \in [0, T]$*, all* $\nu \in \mathcal{P}_2(\mathbb{R}^d)$ *and all* $a \in A$*, the following inequalities hold* $\mathbb{P}$*-a.s.:*

$$\begin{aligned}&\Big(\mathrm{D}\,H_t^\nu(x, y, z, u, a) - \mathrm{D}\,H_t^\nu(x', y', z', u', a) \cdot x - x'\Big)\\&\quad + \Big(b_t^\nu(x, a) - b_t^\nu(x', a) \cdot y - y'\Big) + \Big(\sigma_t^\nu(x, a) - \sigma_t^\nu(x', a) \cdot z - z'\Big)_{\mathrm{Fr}}\\&\quad + \Big(\gamma_t^\nu(x, a) - \gamma_t^\nu(x', a) \cdot u - u'\Big)_{\lambda_t^\mu}\\&\le -\beta_1 \left|x - x'\right|^2 - \beta_2 \Big\{ \left|y - y'\right|^2 + \left\|z - z'\right\|_{\mathrm{Fr}}^2 + \left\|u - u'\right\|_{\lambda_t^\mu}^2 \Big\},\end{aligned}$$

and

$$\left(g^\nu(x) - g^\nu(x') \cdot x - x'\right) \ge \beta_3 \left|x - x'\right|^2,$$

for every

$$(x, y, z, U), (x', y', z', U') \in \mathbb{R}^d \times \mathbb{R}^n \times \mathbb{R}^{d \times k} \times \mathfrak{H}^{\lambda,\mu}$$

on the event $\{\mu_{t-} = \nu,\ U_t = u,\ U_t' = u'\}$.

(iii) On the controls:

(C.8) *The set of admissible actions A is given by a non-empty locally compact subset of a finite-dimensional (Euclidean) space, endowed with its usual topology.*

(C.9) *The set of admissible strategies is given by*

$$\mathcal{A} := \Big\{\alpha : \Omega \times [0, T] \to A \;\Big|\; \alpha \text{ is càdlàg and } \mathbb{F}\text{-adapted, with } \|\alpha\|_{\mathcal{A}} := \sup_{[0,T]} \left(\mathbb{E}\left[|\alpha_t|^8\right]\right)^{\frac{1}{8}} < \infty\Big\}.$$

In order to derive the first and second order variations of the optimally controlled trajectory, we need to analyze the integral term appearing in the usual Taylor's expansion of a smooth function. That is, if $\Psi \in \mathrm{C}^k(V)$, with $k \in \mathbb{N}$ and $V \subset \mathbb{R}^d$ an open and convex set, then

$$\Psi(x) - \sum_{j=0}^{k-1} \frac{1}{j!} \mathrm{D}^j \, \Psi(x_0; x - x_0)$$
$$= \frac{1}{(k-1)!} \int_0^1 \mathrm{D}^k \, \Psi\,(x_0 + \tau(x - x_0); x - x_0) \, \mathrm{d}\,\tau, \tag{3.19}$$

for all x, x_0 in the interior of V, where $\mathrm{D}^j \, \Psi(x_0; x - x_0)$ denotes the j-th differential of Ψ around x_0, in the direction of $x - x_0$.

Remark 3.4

(i) In accordance with the notation introduced earlier—Eq. (3.17)—, the term $\delta \widehat{H}_t^{\mu}\big(\alpha | \widehat{X}, Y, Z, U\big)$ is used to denote the difference of Hamiltonians due to changes away from the optimal control, whenever the adjoint FBSDE is "freezed" at $(\widehat{X}_t, Y_t, Z_t, U_t, M_t)$:

$$\delta \widehat{H}_t^{\mu}\big(\alpha | \widehat{X}, Y, Z, U\big) := H_t^{\mu}\big(\widehat{X}, Y, Z, U, \alpha\big) - H_t^{\mu}\big(\widehat{X}, Y, Z, U, \widehat{\alpha}\big). \tag{3.20}$$

with δH_t^{μ} following the usual difference notation, and

$$\widehat{H}_t^{\mu}(y, z, u) = H(\widehat{\mu}_{t-}, \widehat{X}_t, y, z, u, \widehat{\alpha}_t) \tag{3.21}$$

denotes the evaluation of the admissible Hamiltonian on $(\widehat{X}, \widehat{\alpha})$ given the environment μ and a realization of the backward component (y, z, u, m); see, for instance, (3.3) and (3.17).

(ii) Following from the previous point, we write $\lambda_t^{\mu}(r) \equiv \lambda(r, \mu_{t-})$ in order to signalize the dependence of the environment on the intensity explicitly; contrast this with Remark 2.3((i)). The corresponding transition kernel $K^{\lambda,\mu}$—see (2.5)—is defined in a similar way.

(iii) In addition, we use $(v \cdot \mathrm{M})$ to refer to the vector-matrix multiplication $v^{\top}\mathrm{M}$ whenever the dimensions coincide and the product is well-defined. Recall also that $v^{(k)}$ represents the k-coordinate of v, while $\mathrm{M}^{(i,\cdot)}$ and $\mathrm{M}^{(\cdot,j)}$ are used to denote the i-th row and j-th column of M, respectively.

Observe that the boundedness $\mathrm{d}\,\mathbb{P} \otimes \mathrm{d}\,t$-a.a. on the derivatives and the conditions imposed on the intensity process imply that the Jacobian of H with respect to x is given by

$$\mathrm{D}\, H_t^{\mu}(x, y, z, u, a) = \mathrm{D}\, f_t^{\mu}(x, a) + \big(y \cdot \mathrm{D}\, b_t^{\mu}(x, a)\big) + \sum_{i=1}^{n_1} \Big(z^{(\cdot,i)} \cdot \mathrm{D}\, \sigma_t^{\mu}(x, a)^{(\cdot,i)}\Big)$$
$$+ \sum_{j=1}^{n_0} \Big(u(\cdot)^{(\cdot,j)} \cdot \mathrm{D}\, \gamma_t^{\mu}(\cdot, x, a)^{(\cdot,j)}\Big)_{(\lambda_t^{\mu})^{(j)}}. \tag{3.22}$$

Notice that the last term in the above display involves an integral with respect to the random measure, and equalities should be understood component-wise for $\mathrm{d}\,\mathbb{P}\otimes\mathrm{d}\,t$-a.a. $(\omega, t) \in \Omega \times [0, T]$. Similarly, the Hessian of H is given by

$$
\begin{aligned}
\mathrm{D}^2 H_t^{\mu}(x, y, z, u, a) = &\, \mathrm{D}^2 f_t^{\mu}(x, a) + \Big((y \otimes \mathrm{Id}_d) \cdot \mathrm{D}^2 b_t^{\mu}(x, a) \Big) \\
&+ \sum_{i=1}^{n_1} \Big(\Big(z^{(\cdot,i)} \otimes \mathrm{Id}_d \Big) \cdot \mathrm{D}^2 \sigma_t^{\mu}(x, a)^{(\cdot,i)} \Big) \\
&+ \sum_{j=1}^{n_0} \Big(\big(u(\cdot)^{(\cdot,j)} \otimes \mathrm{Id}_d \big) \cdot \mathrm{D}^2 \gamma_t^{\mu}(\cdot, x, a)^{(\cdot,j)} \Big)_{(\lambda_t^{\mu})^{(j)}},
\end{aligned}
\tag{3.23}
$$

where Id_d represents the d-dimensional identity matrix.

Remark 3.5 Some computations presented in the following require algebraic properties of the Kronecker product, which are recalled next

$$
\begin{aligned}
(\mathrm{Id}_d \otimes v) v_0 = &(\mathrm{Id}_d \otimes v)\mathrm{vec}(v_0^{\top}) = \mathrm{vec}(v v_0^{\top} \mathrm{Id}_d) \\
= &\mathrm{vec}(\mathrm{Id}_d v v_0^{\top}) = (v_0 \otimes \mathrm{Id}_d)\mathrm{vec}(v) = (v_0 \otimes \mathrm{Id}_d) v
\end{aligned}
$$

for $v, v_0 \in \mathbb{R}^d$. They imply, in particular, that for any $M \in \mathbb{R}^{d\times d}$

$$
\begin{aligned}
v_0^{\top} \Big((\mathrm{Id}_d \otimes v) \cdot M v \Big) = & \Big((\mathrm{Id}_d \otimes v)\, v_0 \cdot M v \Big) \\
= & \Big((v_0 \otimes \mathrm{Id}_d)\, v \cdot M v \Big) = v^{\top} \Big((v_0 \otimes \mathrm{Id}_d) \cdot M \Big) v.
\end{aligned}
$$

Let us assume now that $(\widehat{X}, \widehat{\alpha})$ is an optimal pair for the optimization problem (4.35)–(4.36). Then, the *first order adjoint process* is defined as the solution

$$
(Y, Z, U, M) \in \mathcal{S}^2(\mathbb{R}^d) \times \mathbb{H}^2(W) \times \mathbb{H}^2(\tilde{\eta}^{\lambda}) \times \mathcal{H}^{2,\perp} \tag{3.24}
$$

to the following BSDE:

$$
\begin{aligned}
Y_t := &\, \mathrm{D}\,\widehat{g}_T^{\mu\top} + \int_t^T \mathrm{D}\,\widehat{H}_s^{\mu}(Y, Z, U)^{\top}\, \mathrm{d}\,s - \int_t^T Z_s\, \mathrm{d}\,W_s \\
&- \int_{(t,T]\times\mathbf{R}} U_s(r) \tilde{\eta}^{\lambda}(\mathrm{d}\,s, \mathrm{d}\,r) \\
&- (M_T - M_t).
\end{aligned}
\tag{3.25}
$$

Similarly, the *second order adjoint process*

$$
(\widetilde{Y}, \widetilde{Z}, \widetilde{U}, \widetilde{M}) \in \mathcal{S}^2(\mathbb{R}^{d\times d}) \times \mathbb{H}^2(W)^{n_1} \times \mathbb{H}^2(\tilde{\eta}^{\lambda})^{n_0} \times \mathcal{H}^{2,\perp}, \tag{3.26}
$$

with $\widetilde{Z} = (\widetilde{Z}^{(1)}, \ldots, \widetilde{Z}^{(n_1)})$ and $\widetilde{U} = (\widetilde{U}^{(1)}, \ldots, \widetilde{U}^{(n_0)})$, is given as the solution to the BSDE

$$\begin{aligned}\widetilde{Y}_t :=& \mathrm{D}^2 \widehat{g}_T^{\mu} + \int_t^T B_s^{\star\star}\,\mathrm{d}s - \sum_{i=1}^{n_1}\int_t^T \left\{\widetilde{Z}_s^{(i)}\,\mathrm{d}W_s^{(i)} - (\Sigma_s^{\star\star})^{(i)}\,\mathrm{d}s\right\}\\ &- \sum_{j=1}^{n_0}\int_{(t,T]\times\mathbf{R}} \left\{\widetilde{U}_s(r)^{(j)}\widetilde{\eta}^{\lambda}(\mathrm{d}s,\mathrm{d}r)^{(j)} - \Gamma_{s-}^{\star\star}(r)^{(j)}\big(\mathrm{d}s\otimes K_s^{\lambda,\mu}(\mathrm{d}r)^{(j)}\big)\right\}\\ &- (\widetilde{M}_T - \widetilde{M}_t),\end{aligned} \tag{3.27}$$

where $\widetilde{Y}, \widetilde{Z}^{(i)}, \widetilde{U}^{(j)}, \widetilde{M}$ are matrix-valued processes of dimension $d \times d$ for each i and j, while $B^{\star\star}$, $(\Sigma^{\star\star})^{(i)}$ and $\Gamma^{\star\star}(\cdot)^{(j)}$ are defined as

$$B_t^{\star\star} := \mathrm{D}^2 \widehat{H}_t^{\mu}(Y, Z, U) + \mathrm{D}\widehat{b}_t^{\mu\top}\widetilde{Y}_t + \widetilde{Y}_t\,\mathrm{D}\widehat{b}_t^{\mu}, \tag{3.28}$$

$$(\Sigma_t^{\star\star})^{(i)} := \big((\mathrm{D}\widehat{\sigma}_t^{\mu})^{(\cdot,i)}\big)^{\top}\widetilde{Y}_t(\mathrm{D}\widehat{\sigma}_t^{\mu})^{(\cdot,i)} + \big((\mathrm{D}\widehat{\sigma}_t^{\mu})^{(\cdot,i)}\big)^{\top}\widetilde{Z}_t^{(i)} + \widetilde{Z}_t^{(i)}(\mathrm{D}\widehat{\sigma}_t^{\mu})^{(\cdot,i)}, \tag{3.29}$$

$$\begin{aligned}\Gamma_t^{\star\star}(r)^{(j)} :=& \big(\mathrm{D}\widehat{\gamma}_t^{\mu}(r)^{(\cdot,j)}\big)^{\top}\widetilde{Y}_t\,\mathrm{D}\widehat{\gamma}_t^{\mu}(r)^{(\cdot,j)}\\ &+ \big(\mathrm{D}\widehat{\gamma}_t^{\mu}(r)^{(\cdot,j)}\big)^{\top}\widetilde{U}_t(r)^{(j)} + \widetilde{U}_t(r)^{(j)}\,\mathrm{D}\widehat{\gamma}_t^{\mu}(r)^{(\cdot,j)},\end{aligned} \tag{3.30}$$

with the equalities understood component-wise. Then, the *Hamiltonian system for the problem* corresponds to the optimally controlled state trajectory, in conjunction with the *first* and *second order adjoint processes*; in other words, it is defined as the process

$$\big(\widehat{X}, \widehat{Y}, \widehat{Z}, \widehat{U}, \widehat{M}\big) := \big(\widehat{X}, (Y, \widetilde{Y}), (Z, \widetilde{Z}), (U, \widetilde{U}), (M, \widetilde{M})\big), \tag{3.31}$$

such that

- $(\widehat{X}, \widehat{\alpha}) \in \mathcal{S}^2(\mathbb{R}^d) \times \mathcal{A}$ is an optimal pair for (4.35)–(4.36) under the environment μ;
- (Y, Z, U, M), defined as in (3.24), solves the *first order variational equation* (3.25);
- $(\widetilde{Y}, \widetilde{Z}, \widetilde{U}, \widetilde{M})$, defined as in (3.26), solves the *second order variational equation* (3.27).

Observe that the FBSDE system (3.31) fits the setting presented in the previous chapter. The analysis of this system relies on Theorem 2.1. On the other hand, notice that the additional martingale terms M and $\widetilde{M}$ appearing in (3.25) and (3.27), respectively, are a consequence of the multiple, possibly external sources of randomness in the system, and the condition (7) in Definition 2.2. Namely, these processes arise from the martingale decomposition found in the admissible set-up, and are used to compensate the additional information present in the general filtration $\mathbb{F}$. Consequently, as we will see in later sections where the environment itself is endogenous to the system, e.g. when μ coincides with the conditional law of the process given the common noise, these terms automatically disappear.

3.3 Necessary Conditions for Optimality

Now we proceed to find necessary conditions for the pair $(\widehat{X}, \widehat{\alpha})$, which must be satisfied in order to be optimal. These results require some preliminary technical estimations presented next.

3.3.1 A Priori Estimates

Definition 3.1 Given $\epsilon > 0$, let I^ϵ be a sub-interval of $[0, T]$ with Lebesgue measure $|I^\epsilon| = \epsilon$. The ϵ*-variation* (or *spike variation*) *of* $\widehat{\alpha}$ *with respect to* α, denoted by $\alpha^\epsilon \equiv \alpha^{I^\epsilon} \in \mathcal{A}$, is then defined as the (admissible) policy

$$\alpha_t^\epsilon(\omega) := \begin{cases} \alpha_t(\omega), & t \in I^\epsilon, \\ \widehat{\alpha}_t(\omega), & t \in [0, T] \setminus I^\epsilon, \end{cases} \tag{3.32}$$

for all $(\omega, t) \in \Omega \times [0, T]$. Consequently, the ϵ-variational pair is the corresponding admissible pair $(X^\epsilon, \alpha^\epsilon)$

Remark 3.6 The choice for a spike-type variation is due to the fact that the space of admissible actions A is not necessarily convex, and thus, a variation of the form $\widehat{\alpha} + \epsilon v$, for some suitable cadlag process v is not guaranteed to belong to set A. Conversely, if set A were also convex, the following proofs could be simplified by using the convex variational method based on the perturbation $\widehat{\alpha} + \epsilon v$. Particularly, the estimates from Lemma 3.1 can be expressed in terms of this alternative variation process.

Observe that the difference between the optimally controlled state $\widehat{X}$ and its ϵ-variation X^ϵ is described by

$$\begin{aligned} \delta \widehat{X}_t^\epsilon :=& \delta \widehat{X}_0^\epsilon + \int_{(0,t] \cap I^\epsilon} \delta \widehat{b}_s^\mu(X^\epsilon, \alpha) \, \mathrm{d}s + \int_{(0,t] \cap I^\epsilon} \delta \widehat{\sigma}_s^\mu(X^\epsilon, \alpha) \, \mathrm{d}W_s \\ &+ \int_{((0,t] \cap I^\epsilon) \times \mathbf{R}} \delta \widehat{\gamma}_s^\mu(X^\epsilon, \alpha | r) \tilde{\eta}^\lambda(\mathrm{d}s, \mathrm{d}r), \end{aligned} \tag{3.33}$$

with $\delta \widehat{X}_0^\epsilon = X_0^\epsilon - \widehat{X}_0$. Following the same intuition behind [78] and [73] and applying Lemma 1.1, the *first order state variation* of $\delta \widehat{X}^\epsilon$ due to α^ϵ, denoted by ϕ, follows the dynamics

$$\begin{aligned}\phi_t =&\delta \widehat{X}^\epsilon_0 + \int_0^t \mathrm{D}\,\widehat{b}^\mu_s \phi_s \,\mathrm{d}\,s + \sum_{i=1}^{n_1} \int_0^t \Big\{ \big(\delta \widehat{\sigma}^\mu_s(\alpha^\epsilon|\widehat{X})\big)^{(\cdot,i)} + (\mathrm{D}\,\widehat{\sigma}^\mu_s)^{(\cdot,i)}\phi_s \Big\} \,\mathrm{d}\,W^{(i)}_s \\ &+ \sum_{j=1}^{n_0} \int_{(0,t]\times\mathbf{R}} \Big\{ \big(\delta \widehat{\gamma}^\mu_s(\alpha^\epsilon|r,\widehat{X})\big)^{(\cdot,j)} + \mathrm{D}\,\widehat{\gamma}^\mu_s(r)^{(\cdot,j)}\phi_{s-} \Big\}\, \tilde{\eta}^\lambda(\mathrm{d}\,s,\mathrm{d}\,r)^{(j)},\end{aligned} \tag{3.34}$$

for all $t \in [0,T]$, where the equality holds $\mathbb{P}$-a.s. entry-wise, with

$$\begin{aligned}&\delta \widehat{\sigma}^\mu_t(\alpha^\epsilon|\widehat{X}) = \sigma^\mu_t(\widehat{X},\alpha^\epsilon) - \sigma^\mu_t(\widehat{X},\widehat{\alpha}) \\ &\Big(\text{resp. } \delta\widehat{\gamma}^\mu_t(\alpha^\epsilon|r,\widehat{X}) = \gamma^\mu_t(r,\widehat{X},\alpha^\epsilon) - \gamma^\mu_t(r,\widehat{X},\widehat{\alpha})\Big),\end{aligned} \tag{3.35}$$

in accordance to the notation from (3.3), and

$$(\mathrm{D}\,\widehat{\sigma}^\mu_t)^{(\cdot,i)} = \begin{bmatrix} (\mathrm{D}\,\widehat{\sigma}^\mu_t)^{(1,i)} \\ \vdots \\ (\mathrm{D}\,\widehat{\sigma}^\mu_t)^{(d,i)} \end{bmatrix}, \qquad \mathrm{D}\,\widehat{\gamma}^\mu_t(\cdot)^{(\cdot,j)} = \begin{bmatrix} \mathrm{D}\,\widehat{\gamma}^\mu_t(\cdot)^{(1,j)} \\ \vdots \\ \mathrm{D}\,\widehat{\gamma}^\mu_t(\cdot)^{(d,j)} \end{bmatrix},$$

for each $i = 1,\dots,n_1$ and $j = 1,\dots,n_0$, respectively.

Analogously, the *second order variation* ϑ is described as the solution to the SDE

$$\begin{aligned}\vartheta_t =& \int_0^t \Big\{ \delta \widehat{b}^\mu_s(\alpha^\epsilon|\widehat{X}) + (\mathrm{D}\,\widehat{b}^\mu_s)\vartheta_s + \frac{1}{2}\,(\mathrm{Id}_d \otimes \phi_s)^\top\,(\mathrm{D}^2\,\widehat{b}^\mu_s)\phi_s \Big\}\,\mathrm{d}\,s \\ &+ \sum_{i=1}^{n_1}\int_0^t \Big\{ (\mathrm{D}\,\widehat{\sigma}^\mu_s)^{(\cdot,i)}\vartheta_s + \delta\,\mathrm{D}\,\widehat{\sigma}^\mu_s(\alpha^\epsilon|\widehat{X})^{(\cdot,i)}\phi_s \\ &\qquad + \frac{1}{2}(\mathrm{Id}_d \otimes \phi_s)^\top (\mathrm{D}^2\,\widehat{\sigma}^\mu_s)^{(\cdot,i)}\phi_s \Big\}\,\mathrm{d}\,W^{(i)}_s \\ &+ \sum_{j=1}^{n_0}\int_{(0,t]\times\mathbf{R}} \Big\{ \mathrm{D}\,\widehat{\gamma}^\mu_{s-}(r)^{(\cdot,j)}\vartheta_{s-} + \delta\,\mathrm{D}\,\widehat{\gamma}^\mu_{s-}(\alpha^\epsilon|r,\widehat{X})^{(\cdot,j)}\phi_{s-} \\ &\qquad + \frac{1}{2}(\mathrm{Id}_d \otimes \phi_{s-})^\top\,\mathrm{D}^2\,\widehat{\gamma}^\mu_{s-}(r)^{(\cdot,j)}\phi_{s-}\Big\}\tilde{\eta}^\lambda(\mathrm{d}\,s,\mathrm{d}\,r)^{(j)},\end{aligned} \tag{3.36}$$

with $\delta\,\mathrm{D}\,\widehat{\sigma}^\mu_t(\alpha^\epsilon|\widehat{X})$ and $\delta\,\mathrm{D}\,\widehat{\gamma}^\mu_t(\alpha^\epsilon|r,\widehat{X})$ following a similar notation to (3.35), and

$$(\mathrm{D}^2\,\widehat{\sigma}^\mu_t)^{(\cdot,i)} = \begin{bmatrix} (\mathrm{D}^2\,\widehat{\sigma}^\mu_t)^{(1,i)} \\ \vdots \\ (\mathrm{D}^2\,\widehat{\sigma}^\mu_t)^{(d,i)} \end{bmatrix}, \qquad \mathrm{D}^2\,\widehat{\gamma}^\mu_t(\cdot)^{(\cdot,j)} = \begin{bmatrix} \mathrm{D}^2\,\widehat{\gamma}^\mu_t(\cdot)^{(1,j)} \\ \vdots \\ \mathrm{D}^2\,\widehat{\gamma}^\mu_t(\cdot)^{(d,j)} \end{bmatrix},$$

for each $i = 1, \dots, n_1$ and $j = 1, \dots, n_0$, respectively.

Remark 3.7

(i) Observe that from the regularity imposed on the coefficients in Assumption (C.2), the linear SDEs described in (3.34) and (3.36) have a strong solution; see the corresponding discussion on uncoupled linear components in Remark 2.7.

(ii) As in the classical case, notice that when the control does not appear in either the diffusion or the jump terms, we can discard ϕ from them, since it does not contribute to the variation of the control.

The processes ϕ and ϑ above correspond to the stochastic version of the estimates y_1 and y_2 in [73], respectively. From Assumptions B and Theorem 2.1, see also [39], we know the coefficients involved require additional measurability conditions in order to guarantee the existence of both processes.

However, note that these conditions are automatically met once the well-posedness of the controlled state X^α is guaranteed; indeed, due to our standing assumptions on the coefficients, for any given pair (X^α, α) the corresponding drivers of the adjoint BSDEs (Y, Z, U, M) and $(\widetilde{Y}, \widetilde{Z}, \widetilde{U}, \widetilde{M})$ from (3.25) and (3.27), with the pair (X^α, α) a priori fixed, satisfy the conditions for Corollary 4.5 from [39].

Thus, in order to attain the appropriate estimates for the variation of the cost, we require to control the randomness inherent to the environment.

Lemma 3.1 *Let $(\widehat{X}, \widehat{\alpha})$ be an optimal pair, and for $\epsilon > 0$ let $(X^\epsilon, \alpha^\epsilon)$ be the corresponding spike variation with respect to $\alpha \in \mathcal{A}$. Then, for each $k = 1, \dots, 4$,*

$$\sup_{[0,T]} \mathbb{E}\left[|X_t^\alpha|^{2k}\right] \leq O\left((1 + \|\mu\|_{\mathcal{W}} + \|\alpha\|_{\mathcal{A}})^{2k}\right), \tag{3.37}$$

$$\sup_{[0,T]} \mathbb{E}\left[|\delta \widehat{X}_t^\epsilon|^{2k}\right] \leq O\left((\epsilon^{2k} \vee \epsilon^k)(1 + \|\mu\|_{\mathcal{W}} + \|\alpha^\epsilon\|_{\mathcal{A}} + \|\widehat{\alpha}\|_{\mathcal{A}})^{2k}\right), \tag{3.38}$$

$$\sup_{[0,T]} \mathbb{E}\left[|\phi_t|^{2k}\right] \leq O\left(\epsilon^k (1 + \|\mu\|_{\mathcal{W}} + \|\alpha^\epsilon\|_{\mathcal{A}} + \|\widehat{\alpha}\|_{\mathcal{A}})^{2k}\right), \tag{3.39}$$

$$\sup_{[0,T]} \mathbb{E}\left[|\vartheta_t|^{2k}\right] \leq O\left((\epsilon^{2k} \vee \epsilon^{4k})(1 + \|\mu\|_{\mathcal{W}} + \|\alpha^\epsilon\|_{\mathcal{A}} + \|\widehat{\alpha}\|_{\mathcal{A}})^{4k}\right). \tag{3.40}$$

Proof Under Assumptions C, (3.37) is a direct result from the uniformly continuity on locally compact sets and [52, Thm. 7.1], as well as the elementary fact that

$$\left|\sum_{i=1}^n x_i\right|^p \leq n^{p-1} \sum_{i=1}^n |x_i|^p, \qquad p \geq 1,$$

and Gronwall's inequality.

On the other hand, from the Lipschitz condition,

$$\sup_{[0,T]} \mathbb{E}\left[\left|\delta\widehat{b}^{\mu}_t(X^{\epsilon},\alpha^{\epsilon})\right|^k\right] \le O\left(\left(1+\|\mu\|_{\mathcal{W}}+\|\alpha^{\epsilon}\|_{\mathcal{A}}+\|\widehat{\alpha}\|_{\mathcal{A}}\right)^k\right),$$

$$\sup_{[0,T]} \mathbb{E}\left[\left\|\delta\widehat{\sigma}^{\mu}_t(X^{\epsilon},\alpha^{\epsilon})\right\|^{2k}_{\mathrm{Fr}}\right] \le O\left(\left(1+\|\mu\|_{\mathcal{W}}+\|\alpha^{\epsilon}\|_{\mathcal{A}}+\|\widehat{\alpha}\|_{\mathcal{A}}\right)^{2k}\right),$$

$$\sup_{[0,T]} \mathbb{E}\left[\left\|\delta\widehat{\gamma}^{\mu}_t(X^{\epsilon},\alpha^{\epsilon})\right\|^{2k}_{\lambda_t}\right] \le O\left(\left(1+\|\mu\|_{\mathcal{W}}+\|\alpha^{\epsilon}\|_{\mathcal{A}}+\|\widehat{\alpha}\|_{\mathcal{A}}\right)^{2k}\right).$$

Thus, from the boundedness of the derivatives and [52, Thm. 7.1],

$$\mathbb{E}\left|\int_{[0,t]\cap I^{\epsilon}} \delta\widehat{b}^{\mu}_s(X,\alpha)\,\mathrm{d}s\right|^{2k} \le O\left(\epsilon^{2k}\left(1+\|\mu\|_{\mathcal{W}}+\|\alpha^{\epsilon}\|_{\mathcal{A}}+\|\widehat{\alpha}\|_{\mathcal{A}}\right)^{2k}\right),$$

$$\mathbb{E}\left|\int_{[0,t]\cap I^{\epsilon}} \delta\widehat{\sigma}^{\mu}_s(X,\alpha)\,\mathrm{d}W_s\right|^{2k} \le O\left(\epsilon^{k}\left(1+\|\mu\|_{\mathcal{W}}+\|\alpha^{\epsilon}\|_{\mathcal{A}}+\|\widehat{\alpha}\|_{\mathcal{A}}\right)^{2k}\right),$$

$$\begin{aligned}\mathbb{E}&\left|\int_{([0,t]\cap I^{\epsilon})\times\mathbf{R}} \delta\widehat{\gamma}^{\mu}_s(X,\alpha|r)\tilde{\eta}^{\lambda}(\mathrm{d}s,\mathrm{d}r)\right|^{2k}\\ &\le O\left(\epsilon^{k}\left(1+\|\mu\|_{\mathcal{W}}+\|\alpha^{\epsilon}\|_{\mathcal{A}}+\|\widehat{\alpha}\|_{\mathcal{A}}\right)^{2k}\right),\end{aligned}$$

which in turn yields (3.38)–(3.40). □

Proposition 3.1 *Given a random environment μ, let $(X^{\epsilon},\alpha^{\epsilon})$ and $(\widehat{X},\widehat{\alpha})$ be as in Lemma 3.1. Then, for $\epsilon\in(0,1)$,*

$$\sup_{[0,T]} \mathbb{E}\left[\left|\delta\widehat{X}^{\epsilon}_t-\phi_t-\vartheta_t\right|^2\right]=o\left(\epsilon^2\right). \tag{3.41}$$

Proof Using the integral remainder from the Taylor expansion (3.19) on the coefficients, from (3.33)–(3.36) we obtain:

$$\begin{aligned}\delta\widehat{X}^{\epsilon}_t-\phi_t-\vartheta_t =& X^{\epsilon}_t-\overbrace{\left(\widehat{X}_t+\phi_t+\vartheta_t\right)}^{=:\,\rho_t}\\ =&\int_0^t\left\{\delta b^{\mu}_s\left(X^{\epsilon},\rho|\alpha^{\epsilon}\right)+b^{\star}_s\right\}\mathrm{d}s+\int_0^t\left\{\delta\sigma^{\mu}_s\left(X^{\epsilon},\rho|\alpha^{\epsilon}\right)+\sigma^{\star}_s\right\}\mathrm{d}W_s\\ &+\int_{[0,t]\times\mathbf{R}}\left\{\delta\gamma^{\mu}_s\left(X^{\epsilon},\rho|\alpha^{\epsilon},r\right)+\gamma^{\star}_{s-}(r)\right\}\mathrm{d}\tilde{\eta}^{\lambda}(\mathrm{d}s,\mathrm{d}r),\end{aligned}$$

where

$$\begin{aligned}\Psi_t^\star(\cdot) :=& \delta\, \mathrm{D}\,\widehat{\Psi}_t^\mu(\alpha^\epsilon|\cdot, \widehat{X})\vartheta_t + \left(\mathrm{Id}_d \otimes \vartheta_t \cdot \mathrm{D}^2\,\widehat{\Psi}_t^\mu(\cdot)\phi_t\right)\\ &+ \left(\mathrm{Id}_d \otimes (\phi_t + \vartheta_t)\cdot \mathrm{D}^2\,\widehat{\Psi}_t^\mu(\cdot)\vartheta_t\right),\\ &+ \int_0^1 \Big(\mathrm{Id}_d \otimes (\phi_t + \vartheta_t)\cdot \delta\,\mathrm{D}^2\,\widehat{\Psi}_t^\mu\\ &\qquad (\widehat{X} + (1-\tau)(\phi+\vartheta), \alpha^\epsilon|\cdot)\,(\phi_t + \vartheta_t)\Big)\,\mathrm{d}\,\tau,\end{aligned}$$

for $\Psi = b, \sigma, \gamma$. Moreover, from the integral remainder of Taylor's expansion, the identity

$$\delta\Psi_t^\mu\left(X^\epsilon, \rho|\alpha^\epsilon\right) = \int_0^1 \mathrm{D}\,\Psi_t^\mu\left(\rho + (1-\tau)(X^\epsilon - \rho), \alpha^\epsilon\right)(X_t^\epsilon - \rho_t)\,\mathrm{d}\,\tau \tag{3.42}$$

holds for $\mathrm{d}\,\mathbb{P}\otimes\mathrm{d}\,t$-almost all $(\omega, t) \in \Omega \times [0, T]$. Hence, from the boundedness of the derivatives,

$$\left|\delta\Psi_s^\mu\left(X^\epsilon, \rho|\alpha^\epsilon, \cdot\right)\right|^2 \le C\left|X_s^\epsilon - \rho_s\right|^2, \qquad \mathrm{d}\,\mathbb{P}\otimes\mathrm{d}\,s - \text{a.e.}$$

Applying Itô's formula and Gronwall's inequality yields

$$\begin{aligned}\sup_{[0,T]} \mathbb{E}\left[\left|X_t^\epsilon - \rho_t\right|^2\right] \le C \sup_{[0,T]} \mathbb{E}\Bigg[&\left|\int_0^t b_s^\star\,\mathrm{d}\,s\right|^2 + \left|\int_0^t \sigma_s^\star\,\mathrm{d}\,W_s\right|^2\\ &+ \left|\int_{[0,t]\times\mathbf{R}} \gamma_{s-}^\star(r)\,\mathrm{d}\,\tilde{\eta}^\lambda(\mathrm{d}\,s, \mathrm{d}\,r)\right|^2\Bigg].\end{aligned} \tag{3.43}$$

However, from the estimates of Lemma 3.1 and the boundedness of the second order derivatives,

$$\text{r.h.s. of (3.43)} \le O\left(\epsilon^3(1 + \|\mu\|_{\mathcal{W}} + \|\alpha^\epsilon\|_{\mathcal{A}} + \|\widehat{\alpha}\|_{\mathcal{A}})^8\right),$$

thus completing the proof. □

Regarding the cost functional, using the fact that $(\widehat{X}, \widehat{\alpha})$ is optimal, it is clear that

$$\delta\widehat{J}^\mu(X, \alpha) = \mathbb{E}\left[\int_0^T \delta\widehat{f}_t^\mu(X, \alpha)\,\mathrm{d}\,t + \delta\widehat{g}_T^\mu(X)\right] \ge 0 \tag{3.44}$$

for all $(X, \alpha) \in \mathcal{S}^2(\mathbb{R}^d) \times \mathcal{A}$. At the same time, from Assumptions C and previous estimates we have that

$$\mathbb{E}\left[\left|\delta\widehat{g}_T^{\mu}(X^{\epsilon}) - \mathrm{D}\,\widehat{g}_T^{\mu}(\delta\widehat{X}_T^{\epsilon}) - \frac{1}{2}\left(\delta\widehat{X}_T^{\epsilon}\cdot\mathrm{D}^2\,\widehat{g}_T^{\mu}(\delta\widehat{X}_T^{\epsilon})\right)\right|^2\right]$$
$$\leq C\mathbb{E}\left[\left|\delta\widehat{X}_T^{\epsilon}\right|^6\right] = O(\epsilon^3)$$

for all $\epsilon > 0$ sufficiently small, with an analogue result for f. Our goal now is to define the adjoint variational system satisfied by the pay-off J^{μ} and derive the corresponding maximum principle, analogous to (3.6)–(3.9).

3.3.2 Necessary Conditions Through the Hamiltonian System

Theorem 3.1 *Let* $(\widehat{X},\widehat{\alpha})$ *be an optimal pair for the optimization problem* (3.1)–(3.2) *with random environment* μ. *Then, there exists a pair of random processes*

$$(Y, Z, U, M) \in \mathcal{S}^2(\mathbb{R}^d)\times\mathbb{H}^2(W)\times\mathbb{H}^2(\tilde{\eta}^{\lambda})\times\mathcal{H}^{2,\perp} \tag{3.24}$$

and

$$\left(\widetilde{Y},\widetilde{Z},\widetilde{U},\widetilde{M}\right) \in \mathcal{S}^2(\mathbb{R}^{d\times d})\times\mathbb{H}^2(W)^{n_1}\times\mathbb{H}^2(\tilde{\eta}^{\lambda})^{n_0}\times\mathcal{H}^{2,\perp} \tag{3.26}$$

such that

1. (Y, Z, U, M) *is a solution to the first order adjoint BSDE* (3.25)*;*
2. $\left(\widetilde{Y},\widetilde{Z},\widetilde{U},\widetilde{M}\right)$ *is a solution to the second-order adjoint BSDE* (3.27)*;*
3. ***Extended Pontryagin's condition.*** *Recalling* (3.20)*, for any admissible control* $\alpha\in\mathcal{A}$*, the inequality*

$$\delta\widehat{H}_t^{\mu}\left(\alpha\middle|\widehat{X},Y,Z,U\right)+\sum_{i=1}^{n_1}\left(\delta\widehat{\sigma}_t^{\mu}(\alpha|\widehat{X})^{(\cdot,i)}\cdot\frac{\widetilde{Y}_{t-}}{2}\delta\widehat{\sigma}_t^{\mu}(\alpha|\widehat{X})^{(\cdot,i)}\right)_{\mathrm{Fr}}$$
$$+\sum_{j=1}^{n_0}\left(\delta\widehat{\gamma}_t^{\mu}(\alpha|\widehat{X})^{(\cdot,j)}\cdot\frac{\widetilde{Y}_{t-}}{2}\delta\widehat{\gamma}_t^{\mu}(\alpha|\widehat{X})^{(\cdot,j)}\right)_{(\lambda_t^{\mu})^{(j)}}\geq 0 \tag{3.45}$$

holds $\mathrm{d}\,\mathbb{P}\otimes\mathrm{d}\,t$*—a.e.*

Proof Observe that by having the arbitrary random environment μ and the optimal control $\widehat{\alpha}$ fixed, equations (3.2) and (3.25)–(3.27) form a FBSDE in environment-dependent jumps within an admissible set-up. Then, from Theorem 2.1, there exists a Hamiltonian system $(\widehat{X},\widehat{Y},\widehat{Z},\widehat{U},\widehat{M})$ as in (3.31). Hence, we only need to prove (3.45).

Let $0 < \epsilon \ll 1$. From the estimates computed at Lemma 3.1 and Proposition 3.1,

$$\mathbb{E}\left[\delta \widehat{f}_t^{\mu}(X^{\epsilon}, \alpha^{\epsilon})\right] = \mathbb{E}\left[\delta \widehat{f}_t^{\mu}(\rho, \alpha^{\epsilon})\right] + o(\epsilon^2), \quad \mathbb{E}\left[\delta \widehat{g}_T^{\mu}(X^{\epsilon})\right] = \mathbb{E}\left[\delta \widehat{g}_T^{\mu}(\rho)\right] + o(\epsilon^2),$$

where $\rho = \widehat{X} + \phi + \vartheta$. Then,

$$\begin{aligned}
\delta \widehat{J}^{\mu}(\alpha^{\epsilon}, X^{\epsilon}) &= \mathbb{E}\left[\int_0^T \delta \widehat{f}_t^{\mu}(\rho, \alpha^{\epsilon})\,\mathrm{d}t + \delta \widehat{g}_T^{\mu}(\rho)\right] + o(\epsilon^2) \\
&= \mathbb{E}\left[\int_0^T \delta f_t^{\mu}\left(\alpha^{\epsilon}, \widehat{\alpha}\middle|\rho\right)\mathrm{d}t\right] + \mathbb{E}\left[\int_0^T \delta \widehat{f}_t^{\mu}(\rho|\widehat{\alpha})\,\mathrm{d}t + \delta \widehat{g}_T^{\mu}(\rho)\right] + o(\epsilon^2) \\
&= \mathbb{E}\left[\int_0^T \delta f_t^{\mu}\left(\alpha^{\epsilon}, \widehat{\alpha}\middle|\rho\right)\mathrm{d}t\right] + \mathbb{E}\left[\int_0^T \mathrm{D}\,\widehat{f}_t^{\mu}\,(\phi_t + \vartheta_t)\,\mathrm{d}t + \mathrm{D}\,\widehat{g}_T^{\mu}\,(\phi_T + \vartheta_T)\right] \\
&\quad + \frac{1}{2}\mathbb{E}\Big[\int_0^T \Big((\phi_t + \vartheta_t)\cdot \mathrm{D}^2\,\widehat{f}_t^{\mu}(\phi_t + \vartheta_t)\Big)\,\mathrm{d}t \\
&\quad + \Big((\phi_T + \vartheta_T)\cdot \mathrm{D}^2\,\widehat{g}_T^{\mu}(\phi_T + \vartheta_T)\Big)\Big] \\
&\quad + o(\epsilon^2).
\end{aligned} \tag{3.46}$$

Now, from Eqs. (3.22)–(3.23) and Lemma 2.2 we obtain the following identity for the first order variation:

$$\begin{aligned}
&\mathbb{E}\left[\int_0^T \mathrm{D}\,\widehat{f}_t^{\mu}\,(\phi_t + \vartheta_t)\,\mathrm{d}t + \mathrm{D}\,\widehat{g}_T^{\mu}\,(\phi_T + \vartheta_T)\right] \\
&= \mathbb{E}\left[(Y_0 \cdot \delta \widehat{X}_0^{\epsilon})\right] + \mathbb{E}\left[\int_0^T \left\{\delta \widehat{H}_t^{\mu}(\alpha^{\epsilon}|\widehat{X}, Y, Z, U) - \delta \widehat{f}_t^{\mu}(\alpha^{\epsilon}|\widehat{X})\right\}\mathrm{d}t\right] \\
&\quad - \sum_{i=1}^{n_1} \mathbb{E}\left[\int_0^T \left(Z_t^{(i,\cdot)} \cdot \delta\,\mathrm{D}\,\widehat{\sigma}_t^{\mu}(\alpha^{\epsilon}|\widehat{X})^{(\cdot,i)}\phi_t\right)\mathrm{d}t\right] \\
&\quad - \sum_{j=1}^{n_0} \mathbb{E}\left[\int_0^T \left(U_{t-}^{(j,\cdot)} \cdot \delta\,\mathrm{D}\,\widehat{\gamma}_{t-}^{\mu}(\alpha^{\epsilon}|\widehat{X})^{(\cdot,j)}\phi_{t-}\right)_{(\lambda_t^{\mu})^{(j)}}\mathrm{d}t\right] \\
&\quad - \frac{1}{2}\mathbb{E}\left[\int_0^T \left(\phi_t \cdot \left\{\mathrm{D}^2\,\widehat{H}_t^{\mu}(Y, Z, U) - \mathrm{D}^2\,\widehat{f}_t^{\mu}\right\}\phi_t\right)\mathrm{d}t\right].
\end{aligned} \tag{3.47}$$

However, because of the assumptions of boundedness for the derivatives and the square integrability of (Y, Z, U, M), the *a priori* estimates computed in Lemma 3.1 yield

$$\left|\mathbb{E}\left[(Y_0 \cdot \delta \widehat{X}_0^{\epsilon}) + \frac{1}{2}\int_0^T \left(\phi_t \cdot \left\{\mathrm{D}^2\,\widehat{H}_t^{\mu}(Y, Z, U) - \mathrm{D}^2\,\widehat{f}_t^{\mu}\right\}\phi_t\right)\mathrm{d}t\right]\right| \le o(\epsilon).$$

Similarly, for all $1 \le i \le n_1$ and all $1 \le j \le n_0$:

$$\left|\mathbb{E}\left[\int_0^T \left(Z_t^{(i,\cdot)} \cdot \delta\,\mathrm{D}\,\widehat{\sigma}_t^{\mu}(\alpha^\epsilon|\widehat{X})^{(\cdot,i)}\phi_t\right)\mathrm{d}\,t\right]\right|$$
$$+\left|\mathbb{E}\left[\int_0^T \left(U_{t-}^{(j,\cdot)} \cdot \delta\,\mathrm{D}\,\widehat{\gamma}_{t-}^{\mu}(\alpha^\epsilon|\widehat{X})^{(\cdot,j)}\phi_{t-}\right)_{(\lambda_t^\mu)^{(j)}}\mathrm{d}\,t\right]\right| \leq o(\epsilon).$$

Then, plugging (3.47) into (3.46):

$$\begin{aligned}\delta\widehat{J}^{\mu}(X^\epsilon,\alpha^\epsilon) = \mathbb{E}&\left[\int_0^T \left\{\delta f_t^{\mu}\left(\alpha^\epsilon,\widehat{\alpha}\middle|\rho\right) - \delta\widehat{f}_t^{\mu}(\alpha^\epsilon|\widehat{X})\right\}\mathrm{d}\,t\right]\\
&+\frac{1}{2}\mathbb{E}\Big[\int_0^T \left((\phi_t+\vartheta_t)\cdot \mathrm{D}^2\,\widehat{f}_t^{\mu}(\phi_t+\vartheta_t)\right)\mathrm{d}\,t\\
&+\left((\phi_T+\vartheta_T)\cdot \mathrm{D}^2\,\widehat{g}_T^{\mu}(\phi_T+\vartheta_T)\right)\Big]\\
&+\mathbb{E}\left[\int_0^T \delta\widehat{H}_t^{\mu}(\alpha^\epsilon|\widehat{X},Y,Z,U)\,\mathrm{d}\,t\right]+o(\epsilon).\end{aligned} \tag{3.46*}$$

Observe that for the first term at the r.h.s. of (3.46*),

$$\begin{aligned}&\mathbb{E}\left[\int_0^T \left\{\delta f_t^{\mu}\left(\alpha^\epsilon,\widehat{\alpha}\middle|\rho\right) - \delta\widehat{f}_t^{\mu}(\alpha^\epsilon|\widehat{X})\right\}\mathrm{d}\,t\right]\\
=&\mathbb{E}\left[\int_0^T \left\{\delta f_t^{\mu}\left(\rho,\widehat{X}\middle|\alpha^\epsilon\right) - \delta\widehat{f}_t^{\mu}(\rho|\widehat{\alpha})\right\}\mathrm{d}\,t\right]\\
=&\mathbb{E}\Big[\int_0^T \Big\{\delta\,\mathrm{D}\,\widehat{f}_t^{\mu}\left(\alpha^\epsilon\middle|\widehat{X}\right)(\phi_t+\vartheta_t)\\
&\qquad+\frac{1}{2}\left((\phi_t+\vartheta_t)\cdot\delta\,\mathrm{D}^2\,\widehat{f}_t^{\mu}\left(\alpha^\epsilon\middle|\widehat{X}\right)(\phi_t+\vartheta_t)\right)\Big\}\mathrm{d}\,t\Big]+o(\epsilon^2).\end{aligned} \tag{3.48}$$

Likewise, by replicating the proof of Lemma 2.2 twice, first on the product $\mathrm{D}^2\,\widehat{g}_T^{\mu}(\phi_T+\vartheta_T)$ and then on the quadratic form $\big((\phi_T+\vartheta_T)\cdot\mathrm{D}^2\,\widehat{g}_T^{\mu}(\phi_T+\vartheta_T)\big)$, we obtain the following identity for the second order variation:

$$\begin{aligned}&\mathbb{E}\left[\left((\phi_T+\vartheta_T)\cdot\mathrm{D}^2\,\widehat{g}_T^{\mu}(\phi_T+\vartheta_T)\right)\right] = \mathbb{E}\left[(\delta\widehat{X}_0^\epsilon\cdot\tilde{Y}_0\,\delta\widehat{X}_0^\epsilon)\right]\\
&+\int_0^T \mathbb{E}\left[\left((\phi_t+\vartheta_t)\cdot(\tilde{Y}_t-B_t^{\star\star})(\phi_t+\vartheta_t)\right)+\left((\phi_t+\vartheta_t)\cdot\left[\tilde{Y}_t+\tilde{Y}_t^{\top}\right]b_t^{\star\star}\right)\right]\mathrm{d}\,t\\
&-\sum_{i=1}^{n_1}\int_0^T \mathbb{E}\Big[\left((\phi_t+\vartheta_t)\cdot\Sigma_t^{\star\star(i)}(\phi_t+\vartheta_t)\right)-\left((\phi_t+\vartheta_t)\cdot\left[\tilde{Z}_t^{(i)}+\tilde{Z}_t^{(i)\top}\right]\sigma_t^{\star\star(\cdot,i)}\right)\end{aligned}$$

$$
- \left(\sigma_t^{\star\star(\cdot,i)} \cdot \tilde{Y}_t \sigma_t^{\star\star(\cdot,i)} \right) \Big] \mathrm{d}t
$$

$$
- \sum_{j=1}^{n_0} \int_0^T \mathbb{E}\Big[\Big((\phi_t + \vartheta_t) \cdot \Gamma_t^{\star\star(j)} (\phi_t + \vartheta_t) \Big)_{(\lambda_t^\mu)^{(j)}}
$$

$$
- \Big((\phi_t + \vartheta_t) \cdot \left[\tilde{U}_t^{(j)} + \tilde{U}_t^{(j)\top} \right] \gamma_t^{\star\star(\cdot,j)} \Big)_{(\lambda_t^\mu)^{(j)}}
$$

$$
- \left(\gamma_t^{\star\star(\cdot,j)} \cdot \tilde{Y}_t \gamma_t^{\star\star(\cdot,j)} \right)_{(\lambda_t^\mu)^{(j)}} \Big] \mathrm{d}t,
$$

where $B^{\star\star}$, $\Sigma^{\star\star}$ and $\Gamma^{\star\star}$ are defined as in (3.28), (3.29) and (3.30), and $b^{\star\star}$, $\sigma^{\star\star}$ and $\gamma^{\star\star}$ denote the drift, diffusion and jump coefficients of the process $\phi + \theta$, respectively.

Because of the boundedness conditions on the derivatives and the square-integrability of the process $(\widetilde{Y}, \widetilde{Z}, \widetilde{U}, \widetilde{M})$, the linear and quadratic terms on $\phi + \vartheta$ are of order $o(\epsilon^2)$:

$$
\mathbb{E}\Big[\Big((\phi_T + \vartheta_T) \cdot \mathrm{D}^2 \widehat{g}_T^\mu (\phi_T + \vartheta_T) \Big) \Big] = \int_0^T \mathbb{E}\left[(\delta\widehat{\sigma}_t^\mu(\alpha^\epsilon|\widehat{X}) \cdot \widetilde{Y}_t \, \delta\widehat{\sigma}_t^\mu(\alpha^\epsilon|\widehat{X}))_{\mathrm{Fr}} \right] \mathrm{d}t
+ \int_0^T \mathbb{E}\left[(\delta\widehat{\gamma}_t^\mu(\alpha^\epsilon|\widehat{X}) \cdot \widetilde{Y}_t \, \delta\widehat{\gamma}_t^\mu(\alpha^\epsilon|\widehat{X}))_{\lambda_t^\mu} \right] \mathrm{d}t + o(\epsilon^2). \tag{3.49}
$$

Then, from (3.46*), (3.48) and (3.49):

$$
\begin{aligned}
\frac{\delta \widehat{J}^\mu(X^\epsilon, \alpha^\epsilon)}{|\epsilon|} &= \frac{1}{|\epsilon|} \int_0^T \mathbb{E}\Big[\delta \widehat{H}_t^\mu(\alpha^\epsilon|\widehat{X}, Y, Z, U) \Big] \mathrm{d}t \\
&+ \frac{1}{|\epsilon|} \int_0^T \mathbb{E}\Big[\Big(\delta\widehat{\sigma}_t^\mu(\alpha^\epsilon|\widehat{X}) \cdot \frac{\widetilde{Y}_t}{2} \, \delta\widehat{\sigma}_t^\mu(\alpha^\epsilon|\widehat{X}) \Big)_{\mathrm{Fr}} \Big] \mathrm{d}t \\
&+ \frac{1}{|\epsilon|} \int_0^T \mathbb{E}\Big[\Big(\delta\widehat{\gamma}_t^\mu(\alpha^\epsilon|\widehat{X}) \cdot \frac{\widetilde{Y}_t}{2} \, \delta\widehat{\gamma}_t^\mu(\alpha^\epsilon|\widehat{X}) \Big)_{\lambda_t^\mu} \Big] \mathrm{d}t + \frac{o(|\epsilon|)}{|\epsilon|}.
\end{aligned} \tag{3.50}
$$

To guarantee differentiation with respect to ϵ, define $l : [0, T] \times A \to \mathbb{R}$ as

$$
l(t, a) := \mathbb{E}\Big[\delta \widehat{H}_t^\mu(a|\widehat{X}, Y, Z, U) + \Big(\delta\widehat{\sigma}_t^\mu(a|\widehat{X}) \cdot \frac{\widetilde{Y}_t}{2} \, \delta\widehat{\sigma}_t^\mu(a|\widehat{X}) \Big)_{\mathrm{Fr}}
+ \Big(\delta\widehat{\gamma}_t^\mu(a|\widehat{X}) \cdot \frac{\widetilde{Y}_t}{2} \, \delta\widehat{\gamma}_t^\mu(a|\widehat{X}) \Big)_{\lambda_t^\mu} \Big].
$$

From (3.32), (3.44) and (3.50), an application of Lemma 1.1 yields

$$\left.\frac{\mathrm{d}\widehat{J}^{\mu}(X^{\alpha},\alpha)}{\mathrm{d}\alpha}\right|_{\alpha=\widehat{\alpha}} = \lim_{|\epsilon|\to 0^+}\frac{1}{|\epsilon|}\int_{I_\epsilon} l(t,\alpha)\,\mathrm{d}t = \int_0^T l(t,\alpha)\,\mathrm{d}t \geq 0 \tag{3.51}$$

$\mathbb{P}$—a.s. for any arbitrary admissible control $\alpha \in \mathcal{A}$.

Observe that in this last step, we are implicitly using the assumption of the finite dimensionality of the action space A, in order to maintain the "classical" notion of differentiability. Then, using the same arguments as in step (c) of the proof of Theorem 2 in [47] (see also the proof of Theorem 5.1 from [75] or steps 3 and 4 in the proof of Theorem 2.1 from [73]), (3.51) implies (3.45), thus concluding the proof. □

Remark 3.8 To continue with the clarifications between the cases when A is convex or not, see Remark 3.6, when A is convex the steps of the proof, after (3.50), of the above theorem can be simplified by using the variation of the form $\alpha^{\epsilon} := \widehat{\alpha} + \epsilon v$ and noting that the first term of $l(t, \alpha^{\epsilon})$ can be estimated directly by terms of order ϵ, while the last two quadratic terms are of order ϵ^2. Then, the result follows from Lebesgue's differentiation theorem applied to the first term. This is consistent with existing results in literature where the second order BSDE $(\widetilde{Y}, \widetilde{Z}, \widetilde{U}, \widetilde{M})$ is absent from Pontryagin's condition; see the discussion of Cases 1 and 2 presented after Theorem III.3.2 in [78].

3.4 Sufficient Conditions for Optimality

We now complement the results given in Theorem 3.1 by presenting a converse result that describes a set of sufficient conditions that a minimizer must satisfy. Our approach is based on the method adopted in [35].

Theorem 3.2 *Let $(\widehat{X}, \widehat{\alpha}) \in \mathcal{S}^2(\mathbb{R}^d) \times \mathcal{A}$ be an $\widehat{\alpha}$–controlled pair. If*

- *g is a convex function on x;*
- *there exist a process (Y, Z, U, M), defined as in* (3.24)*, that solve the adjoint BSDE* (3.25)*;*
- *the mapping*

$$x \longmapsto \min_{a\in A} H_t^{\mu}(x, Y, Z, U, a) =: \mathcal{H}_t^{\mu}(x), \tag{3.52}$$

  which represents the minimized Hamiltonian at the state x, exists and is convex $\mathrm{d}\mathbb{P}\otimes\mathrm{d}t$–a.e.;
- *for any admissible control $\alpha \in \mathcal{A}$, the inequality (see* (3.20)*)*

$$\delta\widehat{H}_t^{\mu}\left(\alpha\middle|\widehat{X}, Y, Z, U\right) \geq 0 \tag{3.53}$$

  holds $\mathrm{d}\mathbb{P}\otimes\mathrm{d}t$–a.a.;

then, $(\widehat{X}, \widehat{\alpha})$ *is an optimal pair for the control problem in random environment* μ *given by* (3.1)–(3.2).

Remark 3.9 Recall that under our standing assumptions on the coefficients (C.4)–(C.7) the existence of solutions to the adjoint BSDE is guaranteed, as this assumptions correspond to the controlled version of conditions for the well-posedness of FBSDEs from Chap. 2, Theorem 2.1.

Proof Let $\alpha \in \mathcal{A}$ be an arbitrary control, and set $X = X^{\alpha}$. In order to show that (3.44) holds, recall that

$$\delta \widehat{J}^{\mu}(X, \alpha) = \mathbb{E}\left[\int_0^T \delta \widehat{f}_t^{\mu}(X, \alpha)\,\mathrm{d}t + \delta \widehat{g}_T^{\mu}(X)\right].$$

From the convexity of g and [39, Lemma 15],

$$\begin{aligned}
\mathbb{E}\left[\delta \widehat{g}_T^{\mu}(X)\right] \geq & \mathbb{E}\left[\left(\delta \widehat{X}_T \cdot \mathrm{D}\widehat{g}_T^{\mu\top}\right)\right] \\
= & \int_0^T \mathbb{E}\left[\left(\delta \widehat{b}_t^{\mu}(X, \alpha) \cdot Y_t\right) + \left(\delta \widehat{\sigma}_t^{\mu}(X, \alpha) \cdot Z_t\right)_{\mathrm{Fr}}\right]\mathrm{d}t \\
& + \int_0^T \mathbb{E}\left[\left(\delta \widehat{\gamma}_t^{\mu}(X, \alpha) \cdot U_t\right)_{\lambda_t^{\mu}} - \left(\delta \widehat{X}_t \cdot \mathrm{D}\widehat{H}_t^{\mu}(Y, Z, U)^{\top}\right)\right]\mathrm{d}t.
\end{aligned}$$

Hence, from (3.17), (3.52) and the convexity of $\mathcal{H}^{\mu}$,

$$\begin{aligned}
\delta \widehat{J}^{\mu}(X, \alpha) & \geq \int_0^T \mathbb{E}\left[\delta \widehat{H}_t^{\mu}(X, \alpha | Y, Z, U) - \left(\delta \widehat{X}_t \cdot \mathrm{D}\widehat{H}_t^{\mu}(Y, Z, U)^{\top}\right)\right]\mathrm{d}t \\
& \geq \int_0^T \mathbb{E}\left[\delta \widehat{H}_t^{\mu}(X, \alpha | Y, Z, U) - \delta \widehat{\mathcal{H}}_t^{\mu}(X)\right]\mathrm{d}t.
\end{aligned}$$

To conclude, observe that (3.53) implies that

$$H_t^{\mu}(\widehat{X}, Y, Z, U, \widehat{\alpha}) = \min_{a \in A} H_t^{\mu}(\widehat{X}, Y, Z, U, a),$$

and from (3.52),

$$H_t^{\mu}(x, Y, Z, U, \alpha) \geq \mathcal{H}_t^{\mu}(x), \qquad \forall x \in \mathbb{R}^d,$$

$\mathbb{P}$-a.s for all $t \in [0, T]$. As a result,

$$\delta \widehat{H}_t^{\mu}(x, \alpha | Y, Z, U) \geq \delta \widehat{\mathcal{H}}_t^{\mu}(x), \qquad \forall x \in \mathbb{R}^d,$$

therefore completing the proof. □

3.4.1 Simplifications in the Convex Case

We begin by briefly mentioning the relative simplifications that can be made when the space of admissible actions A is convex. In the same vein as with the deterministic case—see Remark 1.1—, our choice for a spike-type variation is due to the fact that the space of admissible actions A is not necessarily convex, and thus, a variation of the form

$$\alpha^\epsilon = \hat{\alpha} + \epsilon v, \quad \text{for some suitable (predictable, integrable) } c\grave{a}dl\grave{a}g \text{ process } v, \tag{3.54}$$

is not guaranteed to belong to set A. Conversely, if set A were also convex, the proofs of Theorems 3.1 and 3.2 could be simplified by using the convex variational method based on (3.54), since the estimates from Lemma 3.1 can then be expressed in terms of this alternative variation process.

Moreover, by using a variation of the form (3.54), the last steps of the proof of Theorem 3.1—after (3.50)—are simplified by noting that the first term of $l(t, \alpha^\epsilon)$ can be estimated directly by terms of order ϵ, while the last two quadratic terms are of order ϵ^2. Then, the result follows from Lebesgue's differentiation theorem applied to the first term. This is consistent with existing results in literature where the second order BSDE $(\widetilde{Y}, \widetilde{Z}, \widetilde{U}, \widetilde{M})$ is absent from Pontryagin's condition; see [78].

3.4.2 Generalization to Infinite-Dimensional Spaces

Regarding possible generalizations, it is possible to extend the results presented in this chapter to infinite dimensional normed spaces. To that end, we first recall the notion of *Fréchet derivative*. For further details, we refer to [27]—from which we also take the notation—and Chapter 34 of [15].

Definition 3.2 Let $F : \mathcal{X} \to \mathcal{Y}$ be a mapping between two normed spaces. The derivative of F at a point $x_0 \in \mathcal{X}$, when it exists, is an element $F'(x_0)$ of $L_C(\mathcal{X}, \mathcal{Y})$—the set of continuous linear mappings from $\mathcal{X}$ to $\mathcal{Y}$—such that

$$\lim_{\substack{x \to x_0 \\ x \neq x_0}} \frac{\left\| F(x) - F(x_0) - F'(x_0)(x - x_0) \right\|_{\mathcal{Y}}}{\|x - x_0\|_{\mathcal{X}}} = 0.$$

We say that F is differentiable at x_0 when such a mapping $F'(x_0)$ exists; in that case, it is unique.

It must be noted that in the above definition, differentiability does not depend on the choice among equivalent norms, and that a function which is differentiable at x_0 is necessarily continuous at x_0. Familiar facts from differential calculus such as *linearity*, the *chain rule*, the *mean-value theorem* and *Fermat's rule* (if F attains a local minimum at a differentiable point x_0, then $F'(x_0) = 0$), all hold for this derivative. If in addition $\mathcal{X}$ is a Hilbert space, the notion of directional derivative is naturally extended to

$$F'(x_0; x) := \lim_{t \downarrow 0} \frac{F(x_0 + tx) - F(x_0)}{t} = 0, \qquad \forall x \in \mathcal{X}.$$

The mapping $x \longmapsto F'(x_0; x)$, referred to as the *Gâteaux derivative* at x_0 in the direction of x, turns out to be linear, with

$$F'(x_0; x) = F'(x_0)x, \qquad \forall x \in \mathcal{X}.$$

Taking these concepts into account, note that the notions of variation introduced above—Eqs. (3.33) through (3.36)—not only are compatible with the differentiability from Definition 3.2, but through the estimate of Proposition 3.1, $\phi + \vartheta$ coincides with the Gâteux derivative of the spike variation process $\delta \hat{X}^{\epsilon}$ with respect to the uniform norm in $L^2(\mathbb{P})$. This means that in the infinite dimensional case, the directional derivatives $\mathrm{D}\,\Psi(x_0; x - x_0)$ in Taylor's expansion (3.19) could be replaced by $\Psi'(x_0; x - x_0)$—see Theorem 34.2 of [15]—, with the inner products replaced by the ones in the corresponding Hilbert space.

Unfortunately, this type of generalization in the state space is not as straightforward as simply "swapping derivatives in the X coordinate" due to the intrinsic relation between the environment and the solution to FBSDEs, as seen earlier in Chap. 2. As an example, take the space $\mathcal{S}_{1,2}(X)$ of *square integrable semimartingales* along with the weaker notion of so-called *vertical derivatives* in the context of *pathwise calculus for non-anticipative functionals* (also *functional Itô calculus*) presented in [5]. In this setting, estimates for the uncoupled forward and backward components already exist in literature and are applicable in our problem—see [5, 28, 42, 58] and the references there-in, just to name a few.

However, their coupling in the continuation argument from Lemma 2.3 relies heavily on the underlying point process and its available stochastic calculus, i.e. its intensity measure. In our problem, dependency on the past can only come from the self-exciting component of the intensity—compare with Sect. 2.4.1—, which is where we impose conditions to control the criticality of the jump process, see [20, 29]. We conjecture that deviating from our setting in this direction is possible; yet, these changes would require a more sophisticated toolkit as well as further specifications on the type of discontinuities, aspects that go beyond the scope of the current text.

Restricting ourselves to Markovian-type of functional spaces without path-dependencies, it is more intuitive—albeit still not trivial—to see how the techniques presented here can be extended. Just in the uncontrolled case with states X taking values in an infinite dimensional space $\mathcal{X}$, the biggest obstacle comes from identifying $\mathbf{G}^{\mu}$ from the G-monotonicity condition (B.6), as it now corresponds to a multilinear operator on a space of the form $\mathcal{X} \times \mathcal{X} \times \mathcal{X}^{n_1} \times \mathfrak{H}^{\lambda(\mathcal{X})}$. A natural candidate would be the dual with respect a modified inner product of $\mathcal{X}$, say $(\cdot\,,\,\cdot)_{G(\mathcal{X})}$; but of course, this requires a more rigorous argument.

For the case of controls over an infinite dimensional set A, the situation is actually simpler, as the differences in (3.35) are by definition equivalent to

$$\delta\widehat{\Psi}^{\mu}_t(\alpha^{\epsilon}|\widehat{X}) = (\widehat{\Psi}^{\mu}_t)' \underbrace{(\alpha_t - \widehat{\alpha}_t)\mathbf{1}_{I^{\epsilon}}}_{=\delta\widehat{\alpha}^{\epsilon}_t} + o(\|\alpha_t - \widehat{\alpha}_t\|^2_{\mathcal{X}})\mathbf{1}_{I^{\epsilon}}$$

whenever the Fréchet derivative at $\widehat{\alpha}_t$, denoted by $(\widehat{\Psi}^{\mu}_t)'$, exists. More formally, instead of (C.5), the next assumption is needed.

(C.5*) **Fréchet differentiability on the control.**

- $(\Psi, \mathrm{D}\,\Psi, \mathrm{D}^2\,\Psi)(\nu, x, \cdot)$ are Fréchet differentiable in A for all $(\nu, x) \in \mathcal{P}_2(\mathbb{R}^d) \times \mathbb{R}^d$, where Ψ can be either b, σ or f;
- $(\gamma, \mathrm{D}\,\gamma, \mathrm{D}^2\,\gamma)(r, \nu, x, \cdot)$ are Fréchet differentiable in A for all $(r, \nu, x) \in \mathbf{R} \times \mathcal{P}_2(\mathbb{R}^d) \times \mathbb{R}^d$.

Then, the first and second order variations, (3.34) and (3.36) respectively, can be rewritten in terms of the derivatives w.r.t. the control variable. In the scalar case, these are simply

$$\begin{aligned}
\phi_t =& \delta\widehat{X}^{\epsilon}_0 + \int_0^t (\mathrm{D}\,\widehat{b}^{\mu}_s)\phi_s\,\mathrm{d}\,s + \int_0^t \Big\{(\widehat{\sigma}^{\mu}_s)'\,\delta\widehat{\alpha}^{\epsilon}_s + (\mathrm{D}\,\widehat{\sigma}^{\mu}_s)\phi_s\Big\}\,\mathrm{d}\,W_s \\
&+ \int_{(0,t]\times\mathbf{R}} \Big\{\big(\widehat{\gamma}^{\mu}_s(r)\big)'\,\delta\widehat{\alpha}^{\epsilon}_{s-} + \mathrm{D}\,\widehat{\gamma}^{\mu}_s(r)\phi_{s-}\Big\}\,\widetilde{\eta}^{\lambda}(\mathrm{d}\,s, \mathrm{d}\,r), \\
\vartheta_t =& \int_0^t \Big\{(\widehat{b}^{\mu}_s)'\,\delta\widehat{\alpha}^{\epsilon}_s + (\mathrm{D}\,\widehat{b}^{\mu}_s)\vartheta_s + \frac{1}{2}(\mathrm{D}^2\,\widehat{b}^{\mu}_s)\phi^2_s\Big\}\,\mathrm{d}\,s \\
&+ \int_0^t \Big\{(\mathrm{D}\,\widehat{\sigma}^{\mu}_s)\vartheta_s + \big(\mathrm{D}\,\widehat{\sigma}^{\mu}_s\big)'\,\delta\widehat{\alpha}^{\epsilon}_s\,\phi_s + \frac{1}{2}(\mathrm{D}^2\,\widehat{\sigma}^{\mu}_s)\phi^2_s\Big\}\,\mathrm{d}\,W_s \\
&+ \int_{(0,t]\times\mathbf{R}} \Big\{\big(\mathrm{D}\,\widehat{\gamma}^{\mu}_{s-}(r)\big)\vartheta_{s-} + \big(\mathrm{D}\,\widehat{\gamma}^{\mu}_{s-}(r)\big)'\,\delta\widehat{\alpha}^{\epsilon}_{s-}\,\phi_{s-} \\
&\qquad + \frac{1}{2}\big(\mathrm{D}^2\,\widehat{\gamma}^{\mu}_{s-}(r)\big)\phi^2_{s-}\Big\}\,\widetilde{\eta}^{\lambda}(\mathrm{d}\,s, \mathrm{d}\,r),
\end{aligned}$$

which are of the same order as before, see Proposition 3.1. Then, following *verbatim* the proofs presented in the chapter, Theorems 3.1–3.2 hold. Furthermore: it is sufficient for the mapping

$$a \longmapsto H^{\mu}_t(a, \widehat{X}, Y, Z, U)$$

to be $\mathbb{P}$—a.s. convex (in the Fréchet sense) at the critical point $\widehat{\alpha}_t$ for the Pontryagin condition (3.53) to be true, hence cementing the idea of the maximum principle as an analog to the classical *second derivative test.*

3.5 A Financial Application: Portfolio Selection with Regime Switching

To close this chapter, we illustrate the use of Theorems 3.1–3.2 by addressing the problem of *Mean-Variance Portfolio Selection within a Regime Switching Market* in Sect. 3.5. This problem is inspired by Example 4.5 in [54], and builds from the example presented in Sect. 2.4.2 of the previous chapter.

Let R and S be the price of a risk-free bond and a risky asset, respectively. Assume that the interest rate b_0 and the excess return rate b_1 offered depend on a classification system ξ that evaluates the current state of the assets on a scale from 1 to n depending on the current variability of the market, represented in this case by a given exogenous environment μ. Mathematically, R and S are defined on an admissible set-up as the solution processes to the (stochastic) differential equations

$$\mathrm{d}\,R_t = b_0(\xi^{\mu}_{t-})R_t\,\mathrm{d}\,t, \qquad \mathrm{d}\,S_t = \left[b_0(\xi^{\mu}_{t-}) + b_1(\xi^{\mu}_{t-})\right] S_t\,\mathrm{d}\,t + \sigma S_t\,\mathrm{d}\,W_t,$$

for all $t \in [0, T]$, where

- $R_0 = 1$ and $S_0 \equiv X_0$ such that $\mu_0 \in \mathcal{P}_2(\mathbb{R})$ and $\sigma \neq 0$,
- b_0 and b_1 are maps from $\mathcal{E} := \{1, \dots, n\}$ to $(0, \infty)$,
- W is a Brownian motion independent of ξ^{μ},
- The regime ξ^{μ} depends to the random environment μ with a finite state space $\mathbf{R}$ and transition probabilities

$$\mathbb{P}\left[\xi^{\mu}_t = j \,\middle|\, \xi^{\mu}_{t-\delta} = i,\, \mu_{t-\delta} = \nu\right] = \begin{cases} \frac{\upsilon_{k_0}(\nu)}{n}\delta + o(\delta), & i \neq j,\\ 1 - (n-1)\frac{\upsilon_{k_0}(\nu)}{n}\delta + o(\delta), & i = j, \end{cases}$$

  where the function υ_{k_0} is defined as

$$\upsilon_{k_0}(\nu) := \frac{m^2_{k_0}(\nu) + 1}{k_0^2 + 1}$$

  for some parameter $k_0 > 0$ and m^p_k as in (2.49).

In other words, ξ^{μ} is a conditional Markov chain in continuous time and transition semigroup

$$K^{(i,j)}(\nu) = \frac{\upsilon_{k_0}(\nu)}{n}\left(1 - n\mathbf{1}_{\{i=j\}}\right), \qquad \forall \nu \in \mathcal{P}_2(\mathbb{R}^d).$$

From our previous discussions on regime-switching diffusions, the chain ξ^{μ} can be rewritten as

$$\xi_t = \int_{\{t\}\times\mathbf{R}} r\, \eta^{\lambda}(\mathrm{d}s, \mathrm{d}r),$$

where η^{λ} is the lifting of an admissible common noise process $(N^{\lambda}, \xi^{\lambda})$ with intensity kernel

$$K^{\lambda,\mu}(t, \mathrm{d}r) = \sum_{j\in\mathbf{R}} K^{(i,j)}(\mu_{t-})\, \delta_{\{j\}}(\mathrm{d}r) = K^{(i,\cdot)}(\mu_{t-})\delta_{\mathbf{R}}(\mathrm{d}r).$$

Consequently, $\big(X_0, \mu, K^{\lambda,\mu}, (N^{\lambda}, \xi), W\big)$ forms an admissible set-up.

Remark 3.10 Observe that in this case ξ^{μ} can be regarded as a Markov chain that "becomes" uniform in their states: the transition probabilities $v_{k_0}(\mu_t)/n$ converge to $1/n$ as the variance of the environment increases. Moreover, it is possible to write the intensity candidate λ explicitly:

$$\lambda_t = \int_{[0,t)\times\mathbb{R}} \psi(r, \mu_{s-}, \xi^{\mu}_{s-})N(\mathrm{d}s, \mathrm{d}r),$$

where N is an independent Poisson random measure on $\mathbb{R}_+ \times \mathbb{R}_+$ and ψ is the integrable, $\mathbb{F}^{\mu,N}$-predictable function

$$\psi(r, \nu, i) := \sum_{j^+=1}^{n-i} j^+\mathbf{1}_{\Gamma^+(\nu,i,j^+)}(r) - \sum_{j^-=1}^{i-1} j^-\mathbf{1}_{\Gamma^-(\nu,i,j^-)}(r),$$

for all $(r, \nu, i) \in \mathbb{R}_+ \times \mathcal{P}_2(\mathbb{R}^d) \times \mathcal{E}$, and $\Gamma^{\pm}$ are defined as

$$\Gamma^+(\nu, i, j) := c_r(\nu) \times \left[(i-1) + \frac{j-1}{n}, (i-1) + \frac{j}{n}\right),$$

$$\Gamma^-(\nu, i, j) := c_r(\nu) \times \left[(i-1) - \frac{j}{n}, (i-1) - \frac{j-1}{n}\right).$$

Let $A = [a_*, a^*] \subset \mathbb{R}$, $a_* < a^*$, be the amount of money available to an investor to put into the risky asset along the time interval $[0, T]$, and let $\mathcal{A}$ be the corresponding set of admissible strategies. Then, for each admissible strategy $\alpha \in \mathcal{A}$, we define the investment cost (i.e., the payoff functional) as the functional

$$J^{\mu}(X, \alpha) := \mathbb{E}\left[\int_0^T \left(X_{t-}^2 - m(\mu_{t-})^2\right) \frac{f(\xi^{\mu}_{t-})}{T}\, \mathrm{d}t + \left(X_{T-}^2 - m(\mu_{T-})^2\right) g(\xi^{\mu}_{T-})\right], \tag{3.55}$$

with $f, g : \mathcal{E} \to (0, \infty)$, and the wealth processes $X \equiv X^\alpha$ follows the controlled SDE

$$X_t = X_0 + \int_0^t \left\{ b_0\left(\xi^\mu_{s-}\right) X_s + b_1\left(\xi^\mu_{s-}\right) \alpha_s \right\} \mathrm{d}s + \sigma \int_0^t \alpha_s \,\mathrm{d}W_s, \tag{3.56}$$

which in turn is well defined due to the compactness of A.

Since b_0, b_1 and g depend on the environment only through the regime ξ, the measurability and Lipschitz conditions hold for the coefficients

$$b^\mu_t(x, a) := b_0\left(\xi^\mu_{t-}\right) x + b_1\left(\xi^\mu_{t-}\right) a, \quad \sigma^\mu_t(x, a) := \sigma a, \quad \gamma^\mu_t(x, a) := 0, \tag{3.57}$$

and

$$f^\mu_t(x, a) := (x^2 - m(\mu_{t-})^2)\frac{f(\xi^\mu_{t-})}{T}, \quad g^\mu_T(x) := (x^2 - m(\mu_{T-})^2) g(\xi^\mu_T). \tag{3.58}$$

Consequently, the path-wise admissible Hamiltonian (3.17) for the problem is given by

$$H^\mu_t(x, y, z, a) = (x^2 - m(\mu_t)^2)\frac{f(\xi^\mu_{t-})}{T} + b_0(\xi^\mu_{t-})xy + \left(b_1(\xi^\mu_{t-})y + \sigma z\right) a, \tag{3.59}$$

with the corresponding adjoint equations

$$\begin{aligned} Y_t =& 2g(\xi^\mu_T)X_{T-} + \int_t^T \left\{ \frac{2f(\xi^\mu_{s-})}{T} X_{s-} + b_0(\xi^\mu_{s-})Y_{s-} \right\} \mathrm{d}s - \int_t^T Z_s \,\mathrm{d}W_s \\ &- \int_{(t,T]\times\mathbb{R}_+} U_s(r)\widetilde{\eta}^\lambda(\mathrm{d}s, \mathrm{d}r) - (M_T - M_t), \end{aligned} \tag{3.60}$$

$$\begin{aligned} \widetilde{Y}_t =& 2g(\xi^\mu_T) + 2\int_t^T \left\{ \frac{f(\xi^\mu_{s-})}{T} + b_0(\xi^\mu_{s-})\widetilde{Y}_{s-} \right\} \mathrm{d}s - \int_t^T \widetilde{Z}_s \,\mathrm{d}W_s \\ &- \int_{(t,T]\times\mathbb{R}_+} \widetilde{U}_s(r)\widetilde{\eta}^\lambda(\mathrm{d}s, \mathrm{d}r) - (\widetilde{M}_T - \widetilde{M}_t). \end{aligned} \tag{3.61}$$

Since the Hamiltonian H is linear on a, the minimization of (3.59) occurs at the extremal points of A. Then, the candidate for optimal control has the form

$$\widehat{\alpha}_t := \widehat{\alpha}(Y_{t-}, Z_{t-}) = \begin{cases} a^*, & \text{if } b_1(\xi^\mu_{t-})Y_{t-} + \sigma Z_{t-} \le 0, \\ a_*, & \text{if } b_1(\xi^\mu_{t-})Y_{t-} + \sigma Z_{t-} > 0, \end{cases} \tag{3.62}$$

and the corresponding $\widehat{\alpha}$-controlled Hamiltonian system (3.56)–(3.60)–(3.61) becomes a Linear FBSDE.

Lastly, note that both g^{μ}_T in (3.58) and the mapping

$$x \longmapsto b_0(\xi^{\mu}_{t-})Y_{t-}x + \big(b_1(\xi^{\mu}_{t-})Y_{t-} + \sigma Z_{t-}\big)\widehat{\alpha}_t$$

are convex $\mathrm{d}t \otimes \mathrm{d}\mathbb{P}$—a.a.; thus, in order to apply Theorem 3.2 we only need to verify the G-monotonicity conditions (C.7), or equivalently, we need to verify Theorem 2.1 from Chap. 2 holds for the Hamiltonian system (3.56)–(3.60)–(3.61). In both instances, we have that the mapping $\mathbf{G}^{\mu}$ from Theorem 2.1 equals

$$\mathbf{G}^{\mu}_t(x, y, z, u) := \begin{bmatrix} -\frac{2f(\xi^{\mu}_t)}{T}x - b_0 y \\ b_0\left(\xi^{\mu}_t\right)x + b_1\left(\xi^{\mu}_t\right)\widehat{\alpha}(y, z) \\ \sigma\widehat{\alpha}(y, z) \\ 0 \end{bmatrix},$$

and therefore

$$\left(\delta\mathbf{G}^{\mu}_t(x, y, z, u)\cdot\delta(x, y, z, u)\right) = -\frac{2f(\xi^{\mu}_{t-})}{T}|\delta x|^2 + \big(b_1(\xi^{\mu}_{t-})\delta y + \sigma\delta z\big)\,\delta\widehat{\alpha}(y, z).$$

However, due to Eq. (3.62),

$$(b_1(i)\delta y + \sigma\delta z)\,\delta\widehat{\alpha}(y, z) \leq 0$$

for all $y, z \in \mathbb{R}$ on the event $\left\{\xi^{\mu}_t = i\right\}$ for all $i \in \mathcal{E}$. Therefore, by taking $\beta_1 = T^{-1}\min_{i\in\mathcal{E}} 2f(i) > 0$, $\beta_2 = 0$ and $\beta_3 = \min_{i\in\mathcal{E}}\{2g(i)\} > 0$, the G-monotonicity holds and the optimality of $\widehat{\alpha}$ in (3.62) is verified.

Chapter 4
Mean-Field Games with Common Noise

Abstract This chapter analyzes the difference between an endogenous and an exogenous random environment, as each has a different nature and interpretation, extending the scope of the models implemented in previous chapters to areas that are very active in research, such as mean-field games. Given a conditional distribution process, described as a measure-valued process depending on the state process and the filtration under which it is conditioned, the corresponding McKean-Vlasov SDE is studied. The connection between this class of stochastic differential equations and mean-field games is established through the phenomenon of chaos propagation and is illustrated by applying these results to the examples presented in the previous chapters.

4.1 Motivation

In order to provide perspective on the results presented in this chapter, we present a natural way to introduce a random environment into a stochastic system, as we pretended to do since the Chap. 2, using the fundamental concept of conditional expectation. We begin by recalling this concept and then proceed to generalize it to stochastic processes.

Given a probability space $(\Omega, \mathcal{F}, \mathbb{P})$ and a Polish space $\mathcal{Y}$, let Y be a random variable mapping Ω into $\mathcal{Y}$ and some sub-σ-field $\mathcal{G}$. As is well known, a regular conditional distribution $\hat{P}(dy|\mathcal{G})(\omega)$ for a random variable Y given some σ-field $\mathcal{G}$ is one of the most typical examples of a random probability measure. This is, for each $\omega \in \Omega$, $\hat{P}(dy|\mathcal{G})(\omega)$ is a probability measure on $\mathcal{Y}$. An important case which motivates the results presented in the following is when X is a random variable that maps $(\Omega, \mathcal{F})$ to a measurable space $(\mathcal{X}, \mathcal{A})$ and $\mathcal{G}$ is generated by X. Then, $\hat{P}(dy|\mathcal{G})(\omega)$ describes the regular conditional distribution of Y given the realization of X. Further, the existing moments of Y are inherited to the regular conditional distribution $\hat{P}(dy|\mathcal{G})(\omega)$. Moreover, when we begin with a measure $\tau = \tau(dx \times dy)$ in the product space $\mathcal{X} \times \mathcal{Y}$, it can be decomposed in terms of the first marginal $\tau_1(A) = \tau(A \times \mathcal{Y})$ and a stochastic kernel $\tau(B|x)$, written as

D. Hernández-Hernández, J. H. Ricalde-Guerrero, *Optimal Control in Random Environments*, SpringerBriefs in Mathematics,
https://doi.org/10.1007/978-3-032-25641-6_4

$$\tau(A \times B) = \int_{A \times B} \tau_1(dx)\tau(dy|x) = \int_A \tau(B|x)\tau_1(dx),$$

for Borel sets A and B in the corresponding σ-fields.

These fundamental results represent the main motivation for extending the results presented for FBSDEs in Chap. 2 to the in-depth study of such systems for some specific random environments. For example, those that depend on the conditional law of some stochastic elements of the model—like one of the noise processes—, motivated by systems with random interaction and processes that are not directly observed. However, the main focus is on mean-field interactions, analyzed in detail, together with propagation of chaos.

To be specific, we will focus on solutions X to the so-called *conditional McKean-Vlasov* SDE with marked jumps

$$\begin{aligned} X_t =& X_0 + \int_0^t b_s(\mathbb{P}^1[X]_{s-}, X_{s-})\,\mathrm{d}\,s + \int_0^t \sigma_s(\mathbb{P}^1[X]_{s-}, X_{s-})\,\mathrm{d}\,W_s \\ &+ \int_{(0,t]\times\mathbf{R}} \gamma_s(r, \mathbb{P}^1[X]_{s-}, X_{s-})\tilde{\eta}^\lambda(\mathrm{d}\,s, \mathrm{d}\,r). \end{aligned} \tag{4.1}$$

Here $\mathbb{P}^1[X]$ represents the conditional measure process

$$\mathbb{P}^1[X] := \left\{\mathbb{P}^1[X]_t := \mathbb{P}\left[\,\cdot\,\middle|\mathcal{F}_T^{(\lambda,N)}\right] \circ (X_t)^{-1},\ t \in [0,T]\right\}, \tag{4.2}$$

which shall be identified with the random environment μ_t introduced in the previous chapters; see (4.12) below for a clearer description of this conditional law process.

These kind of (probability measure-valued) stochastic processes arise naturally as the limit of weakly-interacting particle systems; i.e. systems where the interaction between particles is the same across individuals and decreases when the size of the population grows, yet they remain coupled through the presence of a common source of randomness (a common noise) encoded in their dynamics.

As the number of individuals in the system increases, a *conditional decoupling effect* occurs in which any given particle will start moving independently of the rest, but in general still affected by the randomness produced by the common noise. We call this decoupling effect *propagation of chaos*, and is a key feature of mean-field game and control theory. We refer the reader to [72] for the pioneer historical reference by Sznitman, and to [23] for the two-volume book on the probabilistic aspects of mean-field theory by Carmona and Delarue; for a more relative recent survey on propagation of chaos, we refer to Chaintron and Diez [25].

Physically, our model describes the behavior of a large population of particles inside a box that individually follow a diffusion with certain parameters and only interact with each other through the average position of the whole population; then, at random times an alarm sounds (represented here by the jump times of the marked process), and the box is shaken in a specific way (the corresponding mark), causing a change in the position of the particles.

Note that the box in our analogy is intrinsically linked with the dynamics of the population, and differs from other models in the literature that consider the box itself as a *medium—environment*—determined by an additional source of randomness; see, for example, [64] for a model with continuous McKean-Vlasov dynamics. Here lies the key difference between the model presented here and the one studied in previous chapters, as the conditional law process $\mathbb{P}^1[X]$ described in (4.2) can be regarded as an *endogenous environment* ingrained within the dynamics of the particle X itself.

Nevertheless, it is worth noting that this is not the only way to obtain an endogenous type of random environment. Consider, for instance, the dynamics of a stochastic Stackelberg game with relaxed controls

$$\begin{aligned} X_t =& X_0 + \int_0^t b_s(X_{s-}, \nu_s^L, \nu_s^F)\,\mathrm{d}s + \int_0^t \sigma_s(X_{s-}, \nu_s^L, \nu_s^F)\,\mathrm{d}W_s \\ &+ \int_{(0,t]\times\mathbf{R}} \gamma_s(r, X_{s-}, \nu_{s-}^L, \nu_{s-}^F)\tilde{\eta}^\lambda(\mathrm{d}s, \mathrm{d}r), \end{aligned} \tag{4.3}$$

in which the Leader (respectively, Follower) is tasked to find an $\mathbb{F}^{W,\eta^\lambda}$-adapted ($\mathbb{F}^{W,\eta^\lambda,\nu^F}$-adapted) flow of measures ν^L (ν^F) over some admissible set of actions A, that minimizes some prescribed cost functional $J(\nu^L, \nu^F)$. In this setting, the control problem faced by the Follower can be regarded as one of optimization with an endogenous random environment, where the said environment is provided by the chosen control of the Leader ν^L, who has to disclose his strategy in advance, in the form of a stochastic process in the space of probability measures on the space of actions, conditioned on the information available at each time. More on the topic of Stackelberg game models can be found in the survey paper due to Li and Sethi [49], and references therein.

The probabilistic framework with which we shall work is the same as described in Sect. 2.3.1, with an emphasis on the proposed decomposition of the sample space. This fact turns out to be crucial for the type of random environment we are dealing with in this chapter since, unlike the environments we have studied so far, the measure-valued processes is determined entirely by the elements in the system—to be precise, it is given by the conditional law of a specific process—and cannot be regarded as an external element to the set-up.

Therefore, in order to ensure that the regular conditional probabilities appearing in the next sections are properly defined, as well as allowing us to adequately separate each of the noises affecting the SDEs under study, the details of the construction of this particular probability space and their consequences are discussed in the next section. Of particular relevance is Lemma 4.1 and Proposition 4.1; the former addresses the admissibility of conditional laws as random environments in the context of Sect. 2.3.1—and thus all subsequent results—, while the later is a technical result use in the proof of the propagation of chaos in the second half of the chapter.

4.2 A Technical Digression on the Conditional Law Process

Let

$$(\Omega, \mathcal{F}, \mathbb{P}) := \left(\Omega^0 \times \Omega^1, (\mathcal{F}_T^{(\lambda,N)} \otimes \mathcal{F}_T^W) \vee \mathcal{N}, \mathbb{P}^0 \otimes \mathbb{P}^1\right) \tag{4.4}$$

be the probability space described in the following way. Here, $(\Omega^i, \mathcal{F}^i, \mathbb{P}^i)$, $i = 0, 1$, are probability spaces, with Ω^i being Polish spaces such that:

- N is a n_0-dimensional Poisson processes over $[0, T] \times \mathbf{R} \times \mathbb{R}_+$ with deterministic compensator $\mathrm{d}t \otimes Q(\mathrm{d}r) \otimes \mathrm{d}s$, $(\mathbf{R}, \mathcal{R})$ a Polish space of marks, and Q a n_0-dimensional vector measure with $|Q(\mathbf{R})| < \infty$, see Remark 2.1((ii));
- the function

$$\lambda : \Omega \times [0, T] \times \mathbf{R} \times \mathcal{P}_2(\mathbb{R}^d) \to \mathbb{R}_+^{n_0} \tag{4.5}$$

  is measurable, uniformly continuous in $[0, T]$ and Lipschitz continuous in $\mathcal{P}_2(\mathbb{R}^d)$ (with respect to the $\mathcal{W}_2$-distance), such that λ is an admissible intensity candidate with respect to filtration $\mathbb{F}^{(\lambda,N)} = \left\{\mathcal{F}_t^{(\lambda,N)}, 0 \le t \le T\right\}$ defined as

$$\begin{aligned}\mathcal{F}_t^{(\lambda,N)} :=& \sigma \left\{\lambda_s(r, \nu) : \forall (s, r, \nu) \in [0, t] \times \mathbf{R} \times \mathcal{P}_2(\mathbb{R}^d)\right\} \\ &\vee \sigma \left\{N(B \times C \times D), (B, C, D) \in \mathcal{B}([0, t]) \otimes \mathcal{R} \otimes \mathcal{B}(\mathbb{R}_+)\right\};\end{aligned}$$

- W is a n_1-dimensional Wiener process;
- (λ, N) is defined on $(\Omega^0, \mathcal{F}^0, \mathbb{P}^0)$. W is defined on $(\Omega^1, \mathcal{F}^1, \mathbb{P}^1)$.
- $\mathcal{N}$ denotes the collection of $\mathbb{P}^0 \otimes \mathbb{P}^1$-null sets on the product space.

Similarly, the corresponding filtrations are extended to the product space:

$$\mathbb{F}^0 := \left(\mathbb{F}^{(\lambda,N)} \otimes \left\{\emptyset, \Omega^1\right\}\right) \vee \mathcal{N} = \left\{(\mathcal{F}_t^{(\lambda,N)} \otimes \left\{\emptyset, \Omega^1\right\}) \vee \mathcal{N} =: \mathcal{F}_t^0, 0 \le t \le T\right\}, \tag{4.6}$$

$$\mathbb{F}^1 := \left(\left\{\emptyset, \Omega^0\right\} \otimes \mathbb{F}^W\right) \vee \mathcal{N} = \left\{(\left\{\emptyset, \Omega^0\right\} \otimes \mathcal{F}_t^W) \vee \mathcal{N} =: \mathcal{F}_t^1, 0 \le t \le T\right\}, \tag{4.7}$$

$$\mathbb{F} := \mathbb{F}^0 \vee \mathbb{F}^1. \tag{4.8}$$

Note that $\mathbb{F}$, $\mathbb{F}^0$ and $\mathbb{F}^1$ satisfy the usual conditions, and $\mathbb{F}^0$ and $\mathbb{F}^1$ are independent. Furthermore, $\mathbb{F} = \left(\mathbb{F}^{(\lambda,N)} \otimes \mathbb{F}^W\right) \vee \mathcal{N}$.

For stochastic processes defined on each coordinate space, we abuse slightly of notation and use the same symbols as their natural extension, i.e.

$$\Theta_t(\omega^0, \omega^1) := \Theta_t(\omega^0) \qquad \forall (t, \omega^1) \in [0, T] \times \Omega^1,$$
$$\left(\text{resp. } \Theta_t(\omega^1)\right), \quad \left(\text{resp. } \forall (t, \omega^0) \in [0, T] \times \Omega^0\right). \tag{4.9}$$

We also denote the conditional law of any stochastic process Θ in $\mathbb{R}^d$ given the common noise, as the measured valued process $\mathbb{P}^1[\Theta] = \left\{\mathbb{P}^1[\Theta]_t, t \in [0, T]\right\}$ given by

$$\mathbb{P}^1[\Theta]_t(B) := \mathbb{P}\left[\Theta_t \in B \middle| \mathcal{F}_T^0\right], \qquad \forall B \in \mathcal{B}(\mathbb{R}^d). \tag{4.10}$$

In general, observe that within the current set-up, given a random variable $\Theta : (\Omega, \mathcal{F}) \to \mathcal{X}$, with $\mathcal{X}$ a Polish space, the mapping

$$\Omega^1 \ni \omega^1 \mapsto \Theta(\omega^0, \omega^1)$$

is a random variable on $(\Omega^1, \mathcal{F}^1, \mathbb{P}^1)$ for $\mathbb{P}^0$-almost all $\omega^0 \in \Omega^0$. Define $\mathcal{N}^\Theta \subset \Omega^0$ as the complement of the set in which $\Theta(\omega^0, \cdot)$ is well-defined. Therefore, we can define $\tilde{\mathbb{P}}^1[\Theta] : \Omega^0 \to \mathcal{P}(\mathcal{X})$ as

$$\omega^0 \mapsto \tilde{\mathbb{P}}^1[\Theta] = \begin{cases} \mathbb{P} \circ \Theta(\omega^0, \cdot)^{-1}, & \omega^0 \notin \mathcal{N}^\Theta, \\ \delta_{\{0\}}, & \omega^0 \in \mathcal{N}^\Theta. \end{cases} \tag{4.11}$$

Using [24, Lemma 2.4], this mapping is indeed a $\mathcal{P}(\mathcal{X})$-valued random variable on Ω^0, and coincides with our previous definition of $\mathbb{P}^1[\Theta]$ for random processes in (4.10) $\mathbb{P}^0$-a.s. thanks to the existence of a regular conditional probability measure.

In particular, when Θ is a random variable that takes values in the Polish space of trajectories $\mathrm{D}([0, T]; \mathbb{R}^d)$, then $\mathbb{P}^1[\Theta]$ is a random variable valued at $\mathcal{P}(\mathrm{D}([0, T]; \mathbb{R}^d))$. Since we endowed $\mathcal{P}(\mathrm{D}([0, T]; \mathbb{R}^d))$ with the σ-algebra generated from measurable projections, we have that for each $t \in [0, T]$,

$$\mathbb{P}^1[\Theta]_t = \mathbb{P}^1[\Theta_t], \qquad \mathbb{P}^0 - \text{a.s.}. \tag{4.12}$$

Furthermore, if ΘIQ2 is an $\mathbb{F}$IQ3-adapted stochastic process (respectively, square-integrable with continuous paths), then the corresponding process $\mathcal{P}(\mathbb{R}^d)$IQ4-valued $\mathbb{P}^1[\Theta]$IQ5 is $\mathbb{F}^0$IQ6-adapted (resp. square-integrable with continuous paths) [24, Lemma 2.5]CC1. These facts lead to an intrinsic relation between the discontinuities of both processes Θ and $\mathbb{P}^1[\Theta]$.

Lemma 4.1 *Let Θ be an $\mathbb{R}^d$-valued process with* càdlàg *trajectories. If Θ is square integrable and $\mathbb{F}$-adapted, then there exists a version of $\mathbb{P}^1[\Theta]$ with* càdlàg *trajectories in $\mathcal{P}_2(\mathbb{R}^d)$ (with respect to the 2-Wasserstein distance), $\mathbb{F}^0$-adapted, and its jumping times match those of Θ.*

Proof Observe that for all $t \in [0, T)$ and $\omega^0 \notin \mathcal{N}^\Theta$, where $\mathcal{N}^\Theta \subset \Omega^0$ is a set of null measure,

$$\Theta_t(\omega^0, \omega^1) = \lim_{n\to\infty} \Theta_{t+\frac{1}{n}}(\omega^0, \omega^1), \qquad \forall \omega^1 \in \Omega^1.$$

From our previous discussion, this implies that $\left(\mathbb{P}^1[\Theta]_t\right)$ has $\mathbb{P}^0$-a.s. right-continuous paths in $\mathcal{P}_2(\mathbb{R}^d)$ with respect to the 2-Wasserstein distance. Similarly, by defining

$$\Theta_{t-}(\omega^0, \omega^1) := \lim_{n\to\infty} \Theta_{t-\frac{1}{n}}(\omega^0, \omega^1), \qquad \forall (\omega^0, \omega^1) \in (\Omega^0 \setminus \mathcal{N}^{\tilde{\Theta}}) \times \Omega^1,$$

which we know that exists $\mathbb{P}$-a.s. for all $t \in (0, T]$, and setting

$$\omega^0 \mapsto \tilde{\mathbb{P}}^1[\Theta]_{t-} := \begin{cases} \mathbb{P} \circ (\Theta(\omega^0, \,\cdot\,)_{t-})^{-1}, & \omega^0 \notin \mathcal{N}^\Theta, \\ \delta_{\{0\}}, & \omega^0 \in \mathcal{N}^\Theta, \end{cases}$$

for each $t \in (0, T]$, we have obtained a version of $\left(\mathbb{P}^1[\Theta]_t\right)$ which paths have left limits $\mathbb{P}^0$-a.s. in $\mathcal{P}_2(\mathbb{R}^d)$ with respect to the 2-Wasserstein distance, concluding the proof. □

As mentioned earlier, the previous lemma links back to the definition of random environment introduced at the very beginning of the book and the subsequent notion of admissibility—Definitions 2.1 and 2.2, respectively. It allows us to frame our results *a posteriori*, meaning that we do not necessarily have to account for the presence of a random environment prior to any of the other elements of the set-up but instead we can deduce its presence *endogenously*, through the other established sources of randomness. In terms of mean-field systems, this means that solutions to McKean-Vlasov equations produce admissible set-ups, provided that their solutions exist.

Moving on to our next result, recall from the theory of stochastic integration that, under suitable conditions, strong solutions to SDEs can be regarded as functionals of the underlying noise processes (see, e.g. Theorem 2 in [7] for an instance of SDEs with jumps). This is known in the literature as the *Yamada-Watanabe theorem* due to the original work done by these authors on SDEs driven by a Brownian motion [76]. For McKean-Vlasov type equations this result is technically complicated, in particular when the SDE is conditional to a common noise filtration. However, the construction of the probabilistic model presented in this work allows us to work under the framework established by Carmona and Delarue—Section 1.2.3 in [24]—with relatively minor changes to account for the jumps in the system. Namely, by considering a sufficiently rich input space, all the topological properties required for the corresponding version of the Yamada-Watanabe theorem to hold are satisfied.

Remark 4.1 For the remainder of this section we will assume that a unique strong solution X exists for the conditional McKean-Vlasov equation (4.1). We shall address this existence problem later, in Sect. 4.3.

Let

$$\begin{aligned}\Omega^{\text{input}} :=& \mathbb{R}^d \times \mathrm{D}([0,T];\mathcal{P}_2(\mathbb{R}^d)) \times \mathcal{M}([0,T]\times\mathbf{R})^{n_0}\\ &\times \mathcal{M}_c^*([0,T]\times\mathbf{R}\times\mathbb{R}_+)^{n_0} \times \mathrm{C}\left([0,T];\mathbb{R}^{n_1}\right),\end{aligned} \tag{4.13}$$

such that each marginal is endowed with its corresponding topology. Namely,

- $\mathbb{R}^d$ with the usual Euclidean topology;
- $\mathrm{D}([0,T];\mathcal{P}_2(\mathbb{R}^d))$ with the Skorohod $J1$-topology with respect to the 2-Wasserstein distance $\mathcal{W}_2$ [30];
- $\mathrm{C}([0,T];\mathbb{R}^{n_1})$ with the uniform topology [13];
- $\mathcal{M}([0,T]\times\mathbf{R})$ and $\mathcal{M}_c^*([0,T]\times\mathbf{R}\times\mathbb{R}_+)$ with the weak*-topology [29].

Then, we define

$$(\mathbf{\Omega},\mathcal{G},\mathbf{Q}) = \left(\Omega^{\text{input}}\times\mathrm{D}([0,T];\mathbb{R}^d),\mathcal{B}(\Omega^{\text{input}})\otimes\mathcal{D}([0,T];\mathbb{R}^d),\mathbf{Q}\right) \tag{4.14}$$

as the measurable space with the product σ-algebra of the corresponding Borel σ-algebras and $\mathbf{Q}$ is the probability distribution induced by the random system

$$\mathfrak{C} := \left(X_0, \mathbb{P}^1[X], \int_{\cdot\times\cdot} K_s^\lambda(\mathrm{d}r)\,\mathrm{d}s, N, W, X\right). \tag{4.15}$$

Furthermore, let $\mathbf{G}$ be the augmented filtration generated by

$$\begin{aligned}\mathcal{G}_t :=& \sigma\{X_0\}\vee\sigma\left\{\mathbb{P}^1[X]_s; 0\le s\le t\right\}\\ &\vee\sigma\left\{\int_{B\times C} K_s^\lambda(\mathrm{d}r)\,\mathrm{d}s; B\in\mathcal{B}([0,t]), C\in\mathcal{R}\right\}\\ &\vee\sigma\{N(B\times D); B\in\mathcal{B}([0,t]), D\in\mathcal{R}\otimes\mathcal{B}(\mathbb{R}_+)\}\\ &\vee\sigma\{W_s; 0\le s\le t\}\vee\sigma\{X_s; 0\le s\le t\},\end{aligned}$$

for all $t\in[0,T]$. The reason to choose this space, in particular the selection of $\mathcal{M}([0,T])^{n_0}$ for the coordinate related to the intensity process, is that we can gain some regularity by invoking the Lebesgue differentiation theorem. Indeed, given an admissible kernel K^λ of the form $\lambda_t(r)Q(\mathrm{d}r)\,\mathrm{d}t$, we can recover a regularized version of the kernel by means of the transformation

$$\tilde{K}_t^\lambda(C) := \lim_{n\to\infty} n\int_{\left(t-\frac{1}{n},t\right]\times C}\lambda_s(r)Q(\mathrm{d}r)\,\mathrm{d}s,$$

if the limit exists, and 0 in any other case. Thus, instead of work directly with the intensity candidate λ in the canonical space, we can use the measure $K_t^{\lambda}(\mathrm{d}r)\,\mathrm{d}t$, corresponding to the absolutely continuous compensator of the admissible noise.

Proposition 4.1 *The distribution law induced in* $(\boldsymbol{\Omega}, \mathcal{G}, \mathbf{Q})$ *(see* (4.14)*) by the random element* $\mathfrak{C}$ *defined in* (4.15)*, depends on*

$$\left(X_0, \mathbb{P}^1[X], \int_{\cdot\times\cdot} K_s^{\lambda}(\mathrm{d}r)\,\mathrm{d}s, N, W\right). \tag{4.16}$$

Furthermore, there exists a $\mathcal{B}(\Omega^{\mathrm{input}})/\mathcal{D}([0,T];\mathbb{R}^d)$*-measurable mapping*

$$\boldsymbol{\Phi} : \Omega^{\mathrm{input}} \to \mathrm{D}([0,T];\mathbb{R}^d),$$

depending only on the distribution of the inputs (4.16)*, such that the equality*

$$X = \boldsymbol{\Phi}\left(X_0, \mathbb{P}^1[X], \int_{\cdot\times\cdot} K_t^{\lambda}(\mathrm{d}r)\,\mathrm{d}t, N, W\right)$$

holds $\mathbb{P}$*-a.s.*

Proof Let $\left(\boldsymbol{\chi}, \mathfrak{M}, \boldsymbol{\Lambda}, \mathbf{n}, \mathbf{w}, \mathbf{x}\right)$ be the canonical process on $(\boldsymbol{\Omega}, \mathcal{G}, \mathbf{G}, \mathbf{Q})$ induced by $\mathfrak{C}$. Observe that, by construction, we can follow exactly the same steps as in Theorem 1.33 at [24], setting the backward components equal to zero. Therefore, it is sufficient to verify that $\mathbf{n}$ and $\mathbf{w}$ are mutually independent on the canonical space. Indeed, from the Watanabe and Lévy characterizations, respectively, it is known that the former is a $\mathbf{G}$-Poisson process and the latter is a $\mathbf{G}$-Wiener process. Since both of them are defined on the same filtration where they are already independent, this property is inherited; see Theorem II.6.3 in [41]. □

4.3 Strong Solutions to McKean-Vlasov Equations

Next, we establish the existence of a solution in $(\Omega, \mathcal{F}, \mathbb{P})$ for the McKean-Vlasov SDE (4.1) described at the beginning of the chapter. The following hypotheses are enforced throughout this chapter even without explicit mention, since a particular set of requirements are necessary to treat the endogenous random environment case, in contrast with those assumed for the exogenous one in Chap. 2.

Assumptions D
(i) On the space

(D.1) ***Probability space.*** *The probability space* $(\Omega, \mathcal{F}, \mathbb{P})$ *is defined as in* (4.4)*, equipped with the filtration* $\mathbb{F} = \{\mathcal{F}_t\}$ *given by* (4.8)*, such that an independent* n_0*-dimensional Poisson point processes* N *on* $[0,T] \times \mathbf{R} \times \mathbb{R}_+$ *with intensity*

measure $\mathrm{d}t \otimes Q(\mathrm{d}r) \otimes \mathrm{d}s$ *is supported on* $(\Omega^0, \mathcal{F}^0, \mathbf{P}^0)$*, and an independent* n_1*-dimensional Wiener processes* W *is supported in* $(\Omega^1, \mathcal{F}^1, \mathbb{P}^1)$.

(D.2) ***Initial distribution.*** X_0 *is a* $\mathcal{F}_0^0$*-measurable random variable defined in* $(\Omega^0, \mathcal{F}^0, \mathbb{P}^0)$ *with law* $\nu_0 \in \mathcal{P}_2(\mathbb{R}^d)$ *such that*

$$\mathbb{E}^0\left[\left|X_0^i\right|^2\right] = \int_{\mathbb{R}^d} |x|^2 \, \nu_0(\mathrm{d}x) < \infty. \tag{4.17}$$

(D.3) ***Intensity candidate.*** *The function* λ *is defined as in* (4.5).

(ii) Assumptions on the coefficients *Let*

$$(b, \sigma) : [0, T] \times \mathcal{P}_2(\mathbb{R}^d) \times \mathbb{R}^d \to \mathbb{R}^d \times \mathbb{R}^{d \times n_1},$$

$$\gamma : [0, T] \times \mathbf{R} \times \mathcal{P}_2(\mathbb{R}^d) \times \mathbb{R}^d \to \mathbb{R}^{d \times n_0},$$

be measurable functions, uniformly continuous on $[0, T]$*, such that the following conditions holds.*

(D.4) ***Lipschitz condition.*** *There exists positive constants* $K, K^0 > 0$*, such that for any* $x, x' \in \mathbb{R}^d$*,* $\nu, \nu' \in \mathcal{P}_2(\mathbb{R}^d)$*, the following inequality holds for* $\mathrm{d}\,\mathbb{P}\otimes\mathrm{d}\,t$*-a. a.* $(\omega, t) \in \Omega \times [0, T]$*:*

$$\begin{aligned}
&|\delta b_t(\nu, x)|^2 + \|\delta\sigma_t(\nu, x)\|_{\mathrm{Fr}}^2 \\
&\quad + \int_{\mathbf{R}} \mathrm{Tr}\left(\delta\gamma_t(r, \nu, x)\, \mathrm{diag}\left(\lambda_t(r, \nu) Q(\mathrm{d}r)\right) \delta\gamma_t(r, \nu, x)^{\top}\right) \\
&\leq K^0\, \mathcal{W}_2(\nu, \nu')^2 + K\, |\delta x|^2 .
\end{aligned}$$

(D.5) ***Linear growth.*** *There exists a positive constant* $\beta > 0$ *such that, for* $(\omega, t) \in \Omega \times [0, T]$ *fixed,*

$$\begin{aligned}
&|b_t(\nu, x)| + \|\sigma_t(\nu, x)\|_{\mathrm{Fr}} \\
&\quad + \left(\int_{\mathbf{R}} \mathrm{Tr}\left(\gamma_t(r, \nu, x)\, \mathrm{diag}\left(\lambda_t(\nu, r) Q(\mathrm{d}r)\right) \gamma_t(r, \nu, x)^{\top}\right)\right)^{\frac{1}{2}} \\
&\leq \beta\left\{\left(\int |x|^2 \nu(\mathrm{d}x)\right)^{\frac{1}{2}} + |x|\right\}
\end{aligned}$$

$\mathrm{d}t \otimes \mathrm{d}\mathbb{P}$*-a.e. for* $x \in \mathbb{R}^d$ *and* $\nu \in \mathcal{P}_2(\mathbb{R}^d)$.

(D.6) ***Jump condition.*** *One of the next two conditions is met:*

(D.6.a) λ *does not depend on* ν*:* $\lambda_t(r, \nu) = \lambda_t(r)$ *for all* $\nu \in \mathcal{P}_2(\mathbb{R}^d)$*;*

(D.6.b) the inequalities

$$\left|\gamma_t(\nu, r, x)^{(i,\cdot)}\right|^2 \leq \max_{1\leq j\leq n_0} \left|\lambda_t(\nu, r)^{(j)} - \lambda_t(\nu', r)^{(j)}\right|, \quad i = 1, \ldots, n_0, \tag{4.18}$$

hold for all $(r, \nu, x) \in \mathbf{R} \times \mathcal{P}_2(\mathbb{R}^d) \times \mathbb{R}^d$, $\mathrm{d}\,\mathbb{P} \otimes \mathrm{d}\,t$*-a.e.*

Remark 4.2

(i) Assumption (D.6) above is a technical condition on the jumps that can be interpreted as a trade-off between the intensity and the measure-valued process: On one hand, (D.6.a) corresponds to what can usually be found in the literature of McKean-Vlasov equations with jumps; see, for instance, [1, 32, 33, 53]); on the other hand, (D.6.b) is the result of having a *law-dependent intensity* and can be considered as an alternative—in a marked process setting—to the global Lipschitz condition found in [4], which is also comparable to the notion of *jump diffusion with state-dependent intensity* studied in [6] and [51], for example.
(ii) Observe that, as they are currently written, the assumptions on the coefficients are not dependent on any control. Later, when we invoke these conditions in the controlled case, (D.4)–(D.6) must be understood as valid for every action $a \in A$.

Theorem 4.1 *Under Assumptions D, there exists a unique strong solution process* $X \in \mathcal{S}^2(\mathbb{R}^d)$ *to the conditional McKean-Vlasov SDE with jumps*

$$\begin{aligned} X_t =& X_0 + \int_0^t b_s(\mathbb{P}^1[X]_{s-}, X_{s-})\,\mathrm{d}\,s + \int_0^t \sigma_s(\mathbb{P}^1[X]_{s-}, X_{s-})\,\mathrm{d}\,W_s \\ &+ \int_{(0,t]\times\mathbf{R}} \gamma_s(r, \mathbb{P}^1[X]_{s-}, X_{s-})\tilde{\eta}^\lambda(\mathrm{d}\,s, \mathrm{d}\,r), \end{aligned} \tag{4.1}$$

where η^λ *is the lifting of the marked Poisson process* (N^λ, ξ)*, with* N^λ *a doubly stochastic Poisson process of intensity kernel*

$$K_t^\lambda(\mathrm{d}\,r) = \lambda_t(r, \mathbb{P}^1[X]_{t-})Q(\mathrm{d}\,r),$$

ξ *a sequence of marks on* $(\mathbf{R}, \mathcal{R})$*,* $\tilde{\eta}^\lambda$ *is the compensated martingale measure of* η^λ*, and* W *is an independent Wiener process.*

Proof We now present the proof of the statement under condition (D.6.b), as the proof with condition (D.6.a) follows similar steps. For each $x \in \mathcal{S}^2(\mathbb{R}^d)$ and $\chi \in L^2(\mathcal{F}_0; \mathbb{R}^d)$, define the mapping $\Phi : \mathcal{S}^2(\mathbb{R}^d) \to \mathcal{S}^2(\mathbb{R}^d)$ as

$$\begin{aligned} \Phi(x)_t :=& \chi + \int_0^t b_s\left(\mathbb{P}^1[x]_{s-}, x_{s-}\right)\mathrm{d}\,s + \int_0^t \sigma_s\left(\mathbb{P}^1[x]_{s-}, x_{s-}\right)\mathrm{d}\,W_s \\ &+ \int_{(0,t]\times\mathbf{R}} \gamma_s\left(r, \mathbb{P}^1[x]_{s-}, x_{s-}\right)\tilde{N}\left(\mathrm{d}\,s, \mathrm{d}\,r, \left(0, \lambda_s(r, \mathbb{P}^1[x]_{s-})\right]\right). \end{aligned}$$

Due to the Itô isometry with respect to W and $\tilde{\eta}^\lambda$, as well as Burkholder-Davis-Gundy martingale inequality, the difference $\delta\Phi = \Phi(x) - \Phi(x')$ for any pair $x, x' \in \mathcal{S}^2(\mathbb{R}^d)$ is bounded in norm $\|\cdot\|_*$ by

$$\|\delta\Phi\|_*^2 \le \mathbb{E}\left[\int_0^T \left|\delta b_t(\mathbb{P}^1[x]_{t-}, x_{t-})\right|^2 + \left\|\delta\sigma_t(\mathbb{P}^1[x]_{t-}, x_{t-})\right\|_{\text{Fr}}^2 \mathrm{d}t\right]$$

$$+ \mathbb{E}\Big[\int_0^T \int_{\mathbf{R}} \Big(\delta\gamma_t(r, \mathbb{P}^1[x]_{t-}, x_{t-}) \mathrm{diag}\big(\lambda_t^{\mathbb{P}^1[x]}(r) Q(\mathrm{d}r)\big)$$

$$\cdot\, \delta\gamma_t(r, \mathbb{P}^1[x]_{t-}, x_{t-})\Big)_{\text{Fr}} \mathrm{d}t\Big]$$

$$+ \mathbb{E}\Big[\int_0^T \int_{\mathbf{R}} \Big(\gamma_t^{\mathbb{P}^1[x']}(r, x'_{t-}) \,\mathrm{diag}\left(\left|\delta\lambda_t(r, \mathbb{P}^1[x]_{t-})\right| Q(\mathrm{d}r)\right)$$

$$\cdot\, \gamma_t^{\mathbb{P}^1[x']}(r, x'_{t-})\Big)_{\text{Fr}} \mathrm{d}t\Big],$$

where we have use the notation $\lambda_t^\mu(r)$ from the previous chapter to signal the terms where the measure-valued terms remains constant (see Remark 3.4((ii))).

Observe that the first two integrals at the RHS of the inequality can be bounded above using the Lipschitz condition of the integrands b, σ and γ. For the last term, we make use of (4.18) as follows:

$$\mathbb{E}\Big[\int_0^T \int_{\mathbf{R}} \Big(\gamma_t^{\mathbb{P}^1[x']}(r, x'_{t-}) \,\mathrm{diag}\left(\left|\delta\lambda_t(r, \mathbb{P}^1[x]_{t-})\right| Q(\mathrm{d}r)\right) \cdot \gamma_t^{\mathbb{P}^1[x']}(r, x'_{t-})\Big)_{\text{Fr}} \mathrm{d}t\Big]$$

$$= \sum_{i=1}^{n_0} \mathbb{E}\left[\int_0^T \int_{\mathbf{R}} \left|\gamma_t^{\mathbb{P}^1[x']}(r, x'_{t-})^{(i,\cdot)}\right|^2 \left|\delta\lambda_t(r, \mathbb{P}^1[x]_{t-})^{(i)}\right| Q(\mathrm{d}r)^{(i)} \mathrm{d}t\right]$$

$$\le \sum_{i=1}^{n_0} \mathbb{E}\Big[\int_0^T \int_{\mathbf{R}} \max_{1 \le j \le n_0} \left\{\left|\delta\lambda_t(r, \mathbb{P}^1[x]_{t-})^{(j)}\right|^2\right\} Q^{(i)}(\mathrm{d}r)\, \mathrm{d}t\Big]$$

$$\le n_0\, |Q(\mathbf{R})|\, \mathbb{E}\left[\int_0^T \mathcal{W}_2(\mathbb{P}^1[x]_t, \mathbb{P}^1[x']_t)^2 \,\mathrm{d}t\right],$$

where the last inequality is due to the Lipschitz condition on λ. Then, there exists a constant $C > 0$ (depending on the dimension of the noise processes and the state space but not on x) such that

$$\|\delta\Phi\|_*^2 \le C \int_0^T \left\{\mathbb{E}\left[\mathcal{W}_2(\mathbb{P}^1[x]_t, \mathbb{P}^1[x']_t)^2\right] + \mathbb{E}\left[|\delta x_t|^2\right]\right\} \mathrm{d}t$$

$$\le C \int_0^T \mathbb{E}\left[|\delta x_t|^2\right] \mathrm{d}t,$$

where the last inequality is due to (4.19).

Lastly, for any positive integer $k > 1$, let Φ^k be the k-th composition of Φ with itself. In order to apply Banach's fixed point principle we need to verify that Φ^k is a contracting map for a sufficiently large k. Indeed, by iterating k times we obtain that

$$\begin{aligned}
\left\|\delta\Phi^k\right\|_*^2 &\leq C \int_0^T \mathbb{E}\left[\left|\delta\Phi_t^{k-1}\right|^2\right] \mathrm{d}t \leq C \int_0^T \mathbb{E}\left[\sup_{[0,t]}\left|\delta\Phi_s^{k-1}\right|^2\right] \mathrm{d}t \\
&\leq C^2 \int_0^T \int_0^t \mathbb{E}\left[\left|\delta\Phi_s^{k-2}\right|^2\right] \mathrm{d}s\,\mathrm{d}t = C^2 \int_0^T \frac{T-t}{1} \mathbb{E}\left[\left|\delta\Phi_t^{k-2}\right|^2\right] \mathrm{d}t \\
&\leq C^2 \int_0^T \frac{T-t}{1} \mathbb{E}\left[\sup_{[0,t]}\left|\delta\Phi_s^{k-2}\right|^2\right] \mathrm{d}t \\
&\leq C^3 \int_0^T \frac{T-t}{1} \int_0^t \mathbb{E}\left[\left|\delta\Phi_s^{k-3}\right|^2\right] \mathrm{d}s\,\mathrm{d}t \\
&= C^3 \int_0^T \frac{(T-t)^2}{2!} \mathbb{E}\left[\left|\delta\Phi_t^{k-3}\right|^2\right] \mathrm{d}t \\
&\leq \dots \\
&\leq C^k \int_0^T \frac{(T-t)^{k-1}}{k!} \mathbb{E}\left[|\delta x_t|^2\right] \mathrm{d}t \leq \frac{(CT)^k}{k!} \|\delta x\|_* .
\end{aligned}$$

Thus, by choosing a k such that $k! > (CT)^k$, we find that Φ^k is contractive, completing the proof. □

Before applying the previous result to the solution X of (4.1), recall that from the definition of the Wasserstein distance [22], the inequality

$$\mathcal{W}_2(\nu^1, \nu^2)^2 \leq \mathbb{E}\left[d_{\mathcal{X}}(Y^1, Y^2)^2\right], \tag{4.19}$$

holds for any Y^1, Y^2 independent $\mathcal{X}$-valued random variables defined on the same probability space $(\Omega, \mathcal{F}, \mathbb{P})$ and $\nu^i = \mathbb{P} \circ (Y^i)^{-1}$ for $i = 1, 2$. Then, we obtain the following result.

Corollary 4.1 *Under our standing assumptions, let X be the strong solution process to the conditional McKean-Vlasov equation* (4.1). *Then, the tuple*

$$\left(X_0, \mathbb{P}^1[X], \lambda, (N^\lambda, \xi), W\right)$$

forms a set-up in terms of Definition 2.2, where λ depends on the measure-valued process $\mathbb{P}^1[X]$.

Proof Observe that Conditions 2–7 from Definition 2.2 follow directly from the construction of the probability space presented in Sect. 4.2. In fact, the compatibility condition is obtained from the fact that we consider the product space (4.4)–(4.8), with the corresponding noises being independent.

Hence, we only need to prove that $\mathbb{P}^1[X]$ is a *càdlàg*, square-integrable, $\mathbb{F}$-adapted measure-valued process. Let $c_0 : [0, T] \to \mathbb{R}^d$ be the constant process identically zero. Due to (4.19), it follows that

$$\begin{aligned}\mathcal{W}_{2,d_\infty}\left(\mathbb{P}^1[X], c_0\right)^2 \leq & \int_{D([0,T];\mathbb{R}^d)^2} d_\infty(x, x')^2 \left(\mathbb{P}^1[X] \otimes c_0\right)(\mathrm{d}x, \mathrm{d}x') \\ =& \mathbb{E}^1\left[d_\infty(X, 0)^2\right], \quad \mathbb{P}^0 - a.s.\end{aligned}$$

where d_∞ is the distance induced by the supremum norm. The result is then obtained due to Lemma 4.1 and the square integrability of X. □

Letting X be the strong solution to the system (4.22) established in the above theorem, the pair $\left(\mathbb{P}^1[X], X\right)$ is defined as the *solution process* for the conditional McKean-Vlasov SDE.

4.4 Conditional Propagation of Chaos

In this section we analyze the *propagation of chaos* from the finite population system described next. Consider the class of $\mathbb{R}^d$-valued processes $\left\{X^{i,n}, 1 \leq i \leq n\right\}$, defined as the solution to the *finite-population* SDEs

$$\begin{aligned}X_t^{i,n} =& X_0^i + \int_0^t b_s(\overline{\mu}_{s-}^n, X_{s-}^{i,n})\,\mathrm{d}s + \int_0^t \sigma_s(\overline{\mu}_{s-}^n, X_{s-}^{i,n})\,\mathrm{d}W_s^i \\ &+ \int_{(0,t]\times\mathbf{R}} \gamma_s(r, \overline{\mu}_{s-}^n, X_{s-}^{i,n})\tilde{\eta}^{\lambda,n}(\mathrm{d}s, \mathrm{d}r), \quad t \leq T,\ i = 1, \ldots, n,\end{aligned} \tag{4.20}$$

where $\overline{\mu}^n$ denotes the empirical measure process

$$\overline{\mu}^n = \left\{\overline{\mu}_t^n = \frac{1}{n}\sum_{i=1}^n \delta_{X_t^{i,n}},\ t \in [0, T]\right\}. \tag{4.21}$$

Notice that the elements involved in the description of this system, namely the noise processes, are of the same kind as in the SDE (4.1), and the interaction between the particles is done through the empirical measure.

Then, we are interested in the propagation of chaos of this system, that is, in the convergence of finite population systems to a mean-field limit when the population's size n goes to infinity, and the limit system can be described using a representative particle, with dynamics

$$\begin{aligned} X_t^i = & X_0^i + \int_0^t b_s\left(\mathbb{P}^1\left[X^i\right]_{s-}, X_s^i\right)\mathrm{d}s + \int_0^t \sigma_s\left(\mathbb{P}^1\left[X^i\right]_{s-}, X_s^i\right)\mathrm{d}W_s^i \\ & + \int_{(0,t]\times\mathbf{R}} \gamma_s\left(r, \mathbb{P}^1\left[X^i\right]_{s-}, X_{s-}^i\right)\tilde{\eta}^{\lambda^i}(\mathrm{d}s, \mathrm{d}r), \end{aligned} \tag{4.22}$$

where $\{\mathbb{P}^1\left[X^i\right]_t\}_{t\in[0,T]}$ is the conditional measure process, analogous to the one described in (4.2).

In order to prove this convergence, we first need to extend our assumptions accordingly; i.e., by allowing for a sequence of idiosyncratic noise processes and initial conditions to be defined in the probability space.

Assumptions E
Extended assumptions on the space.

(E.1) ***Probability space.*** *The probability space* $(\Omega, \mathcal{F}, \mathbb{P})$ *is defined as in* (4.4) *such that an independent* n_0*-dimensional Poisson point processes* N *on* $[0, T] \times \mathbf{R}\times\mathbb{R}_+$ *with intensity measure* $\mathrm{d}t\otimes Q(\mathrm{d}r)\otimes\mathrm{d}s$ *is supported on* $(\Omega^0, \mathcal{F}^0, \mathbf{P}^0)$*, and a sequence of independent* n_1*-dimensional Wiener processes* $\{W^i, i \in \mathbb{N}\}$ *is supported in* $(\Omega^1, \mathcal{F}^1, \mathbb{P}^1)$.

(E.2) ***Initial distribution.*** $\{X_0^i, i \in \mathbb{N}\}$ *is a sequence of independent, identically distributed* $\mathcal{F}_0^0$*-measurable random variables defined in* $(\Omega^0, \mathcal{F}^0, \mathbb{P}^0)$ *satisfying* (4.17).

The way we proceed is by considering a *synchronous coupling* between both systems; see [25, 26] and references there-in. That is, by considering a system of conditionally independent, identically distributed conditional McKean-Vlasov equations driven by the same sequence of Brownian motions as in (4.20), as well as an underlying Poisson point process N from which both $\eta^{\lambda,n}$ and η^λ can be obtained; this, in turn, makes it possible to directly compare the two coupled systems. It is easy to see that this conditions are met by virtue of Assumptions (E.1), (E.2) and (D.3), since $\eta^{\lambda,n}$ and η^λ can thus be written as a *thinning*—recall the fundamental properties discussed in Sect. 1.3—of the Poisson point process N:

$$\begin{aligned} \eta^{\lambda,n}(\mathrm{d}t, \mathrm{d}r) &= N(\mathrm{d}t, \mathrm{d}r, (0, \lambda_t(\overline{\mu}_{t-}^n, r)]) \text{ and} \\ \eta^\lambda(\mathrm{d}t, \mathrm{d}r) &= N(\mathrm{d}t, \mathrm{d}r, (0, \lambda_t(\mathbb{P}^1\left[X\right]_{t-}, r)]). \end{aligned} \tag{4.23}$$

Remark 4.3 Throughout this section we assume without explicit mention that Assumptions D hold, and point out again that the probabilistic framework is the one described in Sect. 4.2. Furthermore: observe that from (4.19), it follows that

$$\mathcal{W}_2(\overline{\mu}_x, \delta_{\{x_0\}})^2 \leq \frac{1}{n}\sum_{i=1}^{n}\left|x^i - x_0\right|^2,$$

which together with Assumptions (D.4)–(D.6), imply that (4.20) can be solved using classical results from the theory of SDEs with random coefficients, see Theorems 14.21 and 14.23 [42].

A key concept for the study of the propagation of chaos is the notion of *exchangeability*, which is used to verify the convergence of a symmetric family of SDEs to a McKean-Vlasov system. Recall that $I \subset \mathbb{N}$ is a fixed finite collection of indexes.

Definition 4.1 A family $\{Y^i, i \in I\}$ of random variables is said to be *exchangeable* when the law of $(Y^i)_{i\in I}$ is invariant under any permutation of indexes $i \in I$.

Observe that for independent, identically distributed random variables, exchangeability follows immediately. Guided by this reasoning, the next lemma provides a sufficiently rich setting in which to construct a coupling of exchangeable processes given the information of the jumps.

Lemma 4.2 *For every $i \in I$, let X^i be the solution process to* (4.22). *Then, the family of processes $\left\{X^i, i \in I\right\}$ is conditionally independent, identically distributed given $\mathbb{F}^0$, and thereby exchangeable.*

Proof For each $i \in I$, let $\mu^i := \mathbb{P}^1[X^i]$ and $\mathbb{F}^i = \mathbb{F}^0 \vee \mathbb{F}^{W^i}$. Notice that these coefficients automatically satisfy the measurability and Lipschitz conditions, and hence Theorem 4.1 yields the existence of a strong solution X to the McKean-Vlasov equation (4.22).

Moreover, observe that X^i, defined on $(\Omega, \mathcal{F}, \mathbb{F}^0 \vee \mathbb{F}^{W^i}, \mathbb{P})$, satisfies the conditions of Corollary 4.1. As a result, the tuple

$$\left(X_0^i, \mu^i, \lambda^i, (N^{\lambda^i}, \xi^i), W^i\right) \tag{4.24}$$

with $\lambda^i := \lambda(\cdot, \mu^i, \cdot)$, forms an admissible set-up in terms of Definition 2.2 and the Yamada-Watanabe type of result from Proposition 4.1. In other words, there exists a mapping $\mathbf{\Phi}$, depending only on the law of the set-up, such that

$$\mathbb{P}\left[X^i = \mathbf{\Phi}\left(X_0^i, \mu^i, \lambda^i, (N^{\lambda^i}, \xi^i), W^i\right)\right] = 1, \qquad \forall i \in I.$$

For any pair $(i, j) \in I \times I$, $i \neq j$, let $\underline{X}^{(i,j)}$ be the solution process to the auxiliary SDE

$$\underline{X}_t^{(i,j)} = X_0^i + \int_0^t b_s\left(\mu_{s-}^i, \underline{X}_{s-}^{(i,j)}\right) \mathrm{d}s + \int_0^t \sigma_s\left(\mu_{s-}^i, \underline{X}_{s-}^{(i,j)}\right) \mathrm{d}W_s^j$$
$$+ \int_{[0,t]\times\mathbf{R}} \gamma_s\left(r, \mu_{s-}^i, \underline{X}_{s-}^{(i,j)}\right) \tilde{\eta}^{\lambda^i}(\mathrm{d}s, \mathrm{d}r), \qquad \forall t \in [0, T],$$

The existence of a unique strong solution to this SDE follows from [39, Thm. 3.3], since μ^i is $\mathbb{F}^0$-adapted, and the measurability and Lipschitz conditions for the coefficients are satisfied. Therefore, by the uniqueness in law

$$\mathbb{P}\left[\underline{X}^{(i,j)} = \mathbf{\Phi}\left(X_0^i, \mu^i, \lambda^i, (N^{\lambda^i}, \xi^i), W^j\right)\right] = 1, \qquad \forall i \neq j.$$

Since W^i and W^j are identically distributed on $(\Omega^1, \mathcal{F}^1, \mathbb{P}^1)$, we obtain that for $\mathbb{P}^0$-almost all $\omega^0 \in \Omega^0$ and for all $t \in [0, T]$,

$$\mu_t^i(\omega^0) = \mathbb{P}^1\left[X^i\right]_t(\omega^0) = \mathbb{P}^1\left[\underline{X}^{(i,j)}\right]_t(\omega^0) =: \underline{\mu}_t^{(i,j)}(\omega^0),$$

which in turn yields the equality between intensity candidates

$$\lambda^i = \lambda(\cdot, \mu^i, \cdot) = \lambda(\cdot, \underline{\mu}^{(i,j)}, \cdot) =: \underline{\lambda}^{(i,j)}, \qquad \mathbb{P}^0\text{-a.s.}$$

Since the initial conditions are independent, identically distributed, they are also exchangeable, and hence

$$\mathbb{P}\left[\underline{X}^{(i,j)} = \mathbf{\Phi}\left(X_0^i, \underline{\mu}^{(i,j)}, \underline{\lambda}^{(i,j)}, (N^{\underline{\lambda}^{(i,j)}}, \xi^{(i,j)}), W^j\right)\right]$$
$$= \mathbb{P}\left[\underline{X}^{(i,j)} = \mathbf{\Phi}\left(X_0^j, \underline{\mu}^{(i,j)}, \underline{\lambda}^{(i,j)}, (N^{\underline{\lambda}^{(i,j)}}, \xi^{(i,j)}), W^j\right)\right] = 1;$$

in other words, $\underline{X}^{(i,j)}$ and X^j share the same distribution, thereby proving that

$$\mathbb{P}^1\left[X^i\right]_t(\omega^0) = \mathbb{P}^1\left[X^j\right]_t(\omega^0), \qquad \forall t \in [0, T].$$

for $\mathbb{P}^0$-a.a. $\omega^0 \in \Omega^0$.

Finally, conditional independence of the elements of $\{X^i, i \in I\}$ is verified. Since the underlying spaces are Polish, there exists a regular conditional probability $\mathrm{P} : \Omega^0 \times \mathcal{F} \to [0, 1]$; that is, P is a transition kernel from $(\Omega^0, \mathcal{F}^0)$ into $(\Omega^1, \mathcal{F}^1)$ such that

$$\mathbb{P}\left[(A^0 \times \Omega^1) \cap A\right] = \int_{A^0} \mathrm{P}(\omega^0, A)\, \mathrm{d}\mathbb{P}^0(\omega^0)$$

for all $A^0 \in \mathcal{F}^0$ and all $A \in \mathcal{F}$. However, since $\mathbb{P} = \mathbb{P}^0 \otimes \mathbb{P}^1$,

$$\mathrm{P}(\omega^0, A) = \mathbb{E}^1\left[\mathbf{1}_A(\omega^0, \cdot)\right] = \int_{\Omega^1} \mathbf{1}_A(\omega^0, \omega^1)\,\mathrm{d}\,\mathbb{P}^1(\omega^1).$$

Let $D^i, D^j \in \mathcal{D}([0,T]; \mathbb{R}^d)$. Then, from the definition of conditional expectation,

$$\begin{aligned}
&\int_{A^0\times\Omega^1} \mathbb{E}\left[\mathbf{1}_{\{X^i\in D^i, X^j\in D^j\}}\middle|\mathcal{F}^0\otimes\{\emptyset,\Omega^1\}\right](\omega^0,\omega^1)\,\mathrm{d}\,\mathbb{P}(\omega^0,\omega^1)\\
&= \int_{A^0\times\Omega^1} \mathbf{1}_{\{X^i\in D^i, X^j\in D^j\}}(\omega^0,\omega^1)\,\mathrm{d}\,\mathbb{P}(\omega^0,\omega^1)\\
&= \int_{A^0} \mathrm{P}\left(\omega^0, \left\{X^i\in D^i, X^j\in D^j\right\}\right)\mathrm{d}\,\mathbb{P}^0(\omega^0)\\
&= \int_{A^0} \mathbb{P}^1\left[X^i(\omega^0,\cdot)\in D^i, X^j(\omega^0,\cdot)\in D^j\right]\mathrm{d}\,\mathbb{P}^0(\omega^0)\\
&= \int_{A^0} \mathbb{P}^1\left[X^i(\omega^0,\cdot)\in D^i\right]\mathbb{P}^1\left[X^j(\omega^0,\cdot)\in D^j\right]\mathrm{d}\,\mathbb{P}^0(\omega^0)\\
&= \int_{A^0} \mathrm{P}\left(\omega^0, \left\{X^i\in D^i\right\}\right)\mathrm{P}\left(\omega^0, \left\{X^j\in D^j\right\}\right)\mathrm{d}\,\mathbb{P}^0(\omega^0)\\
&= \int_{A^0\times\Omega^1} \mathbb{E}\left[\mathbf{1}_{\{X^i\in D^i\}}\middle|\mathcal{F}^0\otimes\{\emptyset,\Omega^1\}\right](\omega^0,\omega^1)\\
&\qquad\qquad \times \mathbb{E}\left[\mathbf{1}_{\{X^j\in D^j\}}\middle|\mathcal{F}^0\otimes\{\emptyset,\Omega^1\}\right](\omega^0,\omega^1)\,\mathrm{d}\,\mathbb{P}(\omega^0,\omega^1)
\end{aligned}$$

for all $A^0 \in \mathcal{F}^0$, concluding the proof. □

Finally, we are ready to present the main result of this section:

Theorem 4.2 *The finite population system* $\{X^{i,n}, i \in I, i \le n\}$*, sampled from* (4.20)*, converges to the family of exchangeable McKean-Vlasov differential equations* $\{X^i, i \in I\}$ *defined in Lemma 4.2 in the sense that*

$$\lim_{n\to\infty}\left\{\max_{i\in I}\left\|X^{i,n}-X^i\right\|_*^2 + \max_{i\in I}\sup_{t\in[0,T]}\mathbb{E}\left[\mathcal{W}_2\left(\overline{\mu}_t^{i,n}, \mathbb{P}^1\left[X^i\right]_t\right)^2\right]\right\} = 0. \tag{4.25}$$

Remark 4.4 Observe that due to the Wasserstein inequality (4.19), the first term in the l.h.s. of the above display implies that

$$\lim_{n\to\infty}\sup_{[0,T]}\mathcal{W}_2\left(\mathbb{P}\circ(X_t^{i,n})^{-1}, \mathbb{P}\circ(X_t^i)^{-1}\right) = 0$$

for each $i \in I$; thus, from the relation between Wasserstein distances and weak convergence [69, Thm. 5.11], we conclude that $X^{i,n} \to X^i$ weakly when $n \to \infty$.

Proof Since our main concern is a limit result when $n \to \infty$ and the finite set of indexes $I \subset \mathbb{N}$ is fixed a priori, we assume without loss of generality that $n \geq \max\{i \mid i \in I\}$. We obtain first bounds for the difference of $X^{i,n}$ and X^i. To that end, we recall our difference notation, and write

$$\delta b_t(X^{i,n}, X^i) = b_t\big(\overline{\mu}^n_{t-}, X^{i,n}_{t-}\big) - b_t\big(\mathbb{P}^1\left[X^i\right]_{t-}, X^i_{t-}\big),$$

and similarly for $\delta\sigma$, $\delta\gamma(r,\cdot)$ and $\delta\lambda(r,\cdot)$. We also write $\mathbb{P}^1\left[X^i\right]$ and μ^i interchangeably for the sake of readability in some of the longer equations.

Since both systems $\{X^{i,n}, i \in I\}$ and $\{X^i, i \in I\}$ are driven by the same noise processes (i.e. the same W^i and N, see (4.23)), the Itô Isometry and Burkholder-Davis-Gundy martingale inequality yield

$$\begin{aligned}
&\mathbb{E}\left[\sup_{s\in[0,t]}\left|X^{i,n}_s - X^i_s\right|^2\right] \leq \mathbb{E}\left[\int_0^t \left|\delta b_s(X^{i,n}, X^i)\right|^2 + \left\|\delta\sigma_s(X^{i,n}, X^i)\right\|^2_{\mathrm{Fr}}\,\mathrm{d}s\right]\\
&+\mathbb{E}\left[\int_0^t \int_{\mathbf{R}} \Big(\delta\gamma_s(r, X^{i,n}, X^i)\,\mathrm{diag}\left(\lambda_s(r, \overline{\mu}^n_{s-})Q(\mathrm{d}r)\right)\cdot\delta\gamma_s(r, X^{i,n}, X^i)\Big)_{\mathrm{Fr}}\,\mathrm{d}s\right]\\
&+\mathbb{E}\Bigg[\int_0^t \int_{\mathbf{R}} \Big(\gamma_s(r, \mu^i_{s-}, X^i_{s-})\,\mathrm{diag}\left(\left|\delta\lambda_s(r, X^{i,n}, X^i)\right| Q(\mathrm{d}r)\right)\\
&\qquad\cdot\gamma_s(r, \mu^i_{s-}, X^i_{s-})\Big)_{\mathrm{Fr}}\,\mathrm{d}t\Bigg]
\end{aligned}$$

for all $t \in [0, T]$.

Observe that the r.h.s. of the last inequality can be bounded above using our standing assumptions: for the first two integral terms, the bound is found using the Lipschitz condition on b, γ and σ; for the last integral term, the bound is found by applying one of the Jump conditions from Assumption (D.6), and the Lipschitz condition on λ. Then, there exists a positive constant $C_1 > 0$ such that

$$\mathbb{E}\left[\sup_{s\in[0,t]}\left|X^{i,n}_s - X^i_s\right|^2\right] \leq C_1 \int_0^t \left\{\mathbb{E}\left[\mathcal{W}_2(\overline{\mu}^n_s, \mu^i_s)^2\right] + \mathbb{E}\left[\left|X^{i,n}_s - X^i_s\right|^2\right]\right\}\mathrm{d}s \tag{4.26}$$

for all $i \in I$ and all $t \in [0, T]$.

For all $j \in \{1, \dots, n\} \setminus I$, let X^j be a strong solution to the conditional McKean-Vlasov equation (4.22) with respect to the Brownian motion W^j and write $\overline{\mu} := n^{-1}\sum_{i=1}^n \delta_{X^i}$. Then, for ω^0 fixed,

$$\mathbb{E}^1\left[\mathcal{W}_2(\overline{\mu}^n_t, \mu^i_t)^2\right] \leq \mathbb{E}^1\left[\mathcal{W}_2(\overline{\mu}^n_t, \overline{\mu}_t)^2\right] + \mathbb{E}^1\left[\mathcal{W}_2(\overline{\mu}_t, \mu^i_t)^2\right]$$

$$\leq \frac{1}{n}\sum_{i=1}^{n}\mathbb{E}^1\left[\left|X_t^{i,n}-X_t^i\right|^2\right]+\mathbb{E}^1\left[\mathcal{W}_2(\overline{\mu}_t,\mu_t^i)^2\right]$$
$$=\mathbb{E}^1\left[\left|X_t^{i,n}-X_t^i\right|^2\right]+\mathbb{E}^1\left[\mathcal{W}_2(\overline{\mu}_t,\mu_t^i)^2\right],$$

where the last equality is due to the conditional exchangeability of $\left\{X^i, 1 \leq i \leq n\right\}$ (Lemma 4.2), and thus, the conditional exchangeability of $\{(X^{i,n}, X^i), 1 \leq i \leq n\}$. Moreover, since $\left\{X^i, 1 \leq i \leq n\right\}$ are conditionally independent, identically distributed given $\mathcal{F}^0$, from Glivenko-Cantelli theorem (see Theorem 5.8 at [23]) it follows that for all $i \in I$

$$\mathbb{P}^0\left[\lim_{n\to\infty}\mathbb{E}^1\left[\mathcal{W}_2\left(\overline{\mu}_t,\mathbb{P}^1\left[X^i\right]_t\right)^2\right]=0\right]=1, \qquad \forall t\in[0,T]. \tag{4.27}$$

However, due to (4.19) and the triangle inequality,

$$\mathbb{E}^1\left[\mathcal{W}_2\left(\overline{\mu}_t,\mathbb{P}^1\left[X^i\right]_t\right)^2\right]\leq\frac{2}{n}\sum_{j=1}^{n}\mathbb{E}^1\left[\left|X_t^j\right|^2\right]+2\mathbb{E}^1\left[\left|X_t^i\right|^2\right]=4\mathbb{E}^1\left[\left|X_t^i\right|^2\right]. \tag{4.28}$$

Given that $\max_{i\in I}\mathbb{E}\left[\sup_{t\in[0,T]}\left|X_t^i\right|^2\right]<\infty$ (which holds since I is fixed and each X^i is well-posed), by Dominated Convergence we deduce that

$$\lim_{n\to\infty}\mathbb{E}\left[\mathcal{W}_2\left(\overline{\mu}_t,\mathbb{P}^1\left[X^i\right]_t\right)^2\right]=0, \qquad \forall t\in[0,T]. \tag{4.29}$$

To verify that the convergence is uniform, let $s,t \in [0,T]$. From Cauchy-Schwartz and (4.28) we get

$$\left|\mathbb{E}\left[\mathcal{W}_2\left(\overline{\mu}_t,\mu_t^i\right)^2\right]-\mathbb{E}\left[\mathcal{W}_2\left(\overline{\mu}_s,\mu_s^i\right)^2\right]\right|^2$$
$$\leq\mathbb{E}\left[\left\{\mathcal{W}_2\left(\overline{\mu}_t,\mu_t^i\right)+\mathcal{W}_2\left(\overline{\mu}_s,\mu_s^i\right)\right\}^2\right]\mathbb{E}\left[\left\{\mathcal{W}_2\left(\overline{\mu}_t,\mu_t^i\right)-\mathcal{W}_2\left(\overline{\mu}_s,\mu_s^i\right)\right\}^2\right]$$
$$\leq 16\left\|X^i\right\|_*^2\mathbb{E}\left[\left\{\mathcal{W}_2\left(\overline{\mu}_t,\mu_t^i\right)-\mathcal{W}_2\left(\overline{\mu}_s,\mu_s^i\right)\right\}^2\right]$$
$$\leq 32\left\|X\right\|_*^2\left\{\mathbb{E}\left[\mathcal{W}_2\left(\overline{\mu}_t,\overline{\mu}_s\right)^2\right]+\mathbb{E}\left[\mathcal{W}_2\left(\mu_t^i,\mu_s^i\right)^2\right]\right\}$$
$$\leq 64\left\|X\right\|_*^2\mathbb{E}\left[\left|X_t^i-X_s^i\right|^2\right],$$

where the last inequality is due to (4.19) and the conditional exchangeability of $\{X^i, 1 \leq i \leq n\}$. Then, due to the square-integrability of the *càdlàg* processes X^i and $\mathbb{P}^1\left[X^i\right]$, and the boundedness of the coefficients on bounded sets of $[0,T] \times \mathcal{P}_2(\mathbb{R}^d) \times \mathbb{R}^d$, there exists a positive constant C' such that

$$\left|\mathbb{E}\left[\mathcal{W}_2\left(\overline{\mu}_t, \mathbb{P}^1\left[X^i\right]_t\right)^2\right] - \mathbb{E}\left[\mathcal{W}_2\left(\overline{\mu}_s, \mathbb{P}^1\left[X^i\right]_s\right)^2\right]\right|^2 \leq C' \left|t-s\right|.$$

Therefore, by equicontinuity, the convergence in (4.29) is uniform on t:

$$\lim_{n\to\infty} \sup_{t\in[0,T]} \mathbb{E}\left[\mathcal{W}_2\left(\overline{\mu}_t, \mathbb{P}^1\left[X^i\right]_t\right)^2\right] = 0, \qquad \forall i \in I. \tag{4.30}$$

Going back to (4.26), from Gronwall's inequality we obtain that

$$\mathbb{E}\left[\sup_{s\in[0,t]} \left|X_s^{i,n} - X_s^i\right|^2\right] \leq C \sup_{t\in[0,T]} \mathbb{E}\left[\mathcal{W}_2\left(\overline{\mu}_t, \mathbb{P}\circ (X_t^i)^{-1}\right)^2\right],$$

where C is a positive constant dependent on T. Then, from (4.30),

$$\lim_{n\to\infty} \mathbb{E}\left[\sup_{s\in[0,t]} \left|X_s^{i,n} - X_s^i\right|^2\right] = 0, \qquad \forall t \in [0,T], \forall i \in I, \tag{4.31}$$

thereby obtaining the first term in the asymptotic limit of (4.25). Lastly, for the second term, note that

$$\begin{aligned}\sup_{t\in[0,T]} \mathbb{E}\left[\mathcal{W}_2\left(\overline{\mu}_t^n, \mu_t^i\right)^2\right] &\leq \underbrace{2 \sup_{t\in[0,T]} \mathbb{E}\left[\frac{1}{n}\sum_{i=1}^n \left|X_t^{i,n} - X_t^i\right|^2\right]}_{\to 0 \ (4.31)} \\ &\quad \underbrace{+2 \sup_{t\in[0,T]} \mathbb{E}\left[\mathcal{W}_2\left(\overline{\mu}_t, \mu_t^i\right)^2\right]}_{\to 0 \ (4.30)},\end{aligned}$$

completing the proof. □

Remark 4.5 Notice that the convergence of the empirical law $\overline{\mu}^n$ to the conditional law, instead of convergence to a deterministic limit as in the classical case (the usual law of large numbers), is due to the presence of a common noise in the system since the set-ups (4.24) are only *conditionally* independent and identically distributed (recall the discussion at the start of Sect. 4.2). The use of Glivenko-Cantelli theorem in (4.27) guarantees the convergence to an $\mathcal{F}^0$-measurable distribution $\mathbb{P}^1[X]$;

heuristically, this can be seen as a deterministic limit law given that the common noise is known.

4.4.1 An Application to Regime Switching Diffusions

To finalize this section, we present an example based on the model of a regime switching diffusion presented throughout the previous chapters. More precisely, consider the setting described in Sect. 2.4.2: Let $\mathbf{R} \subset \mathbb{R}^n$ be a finite set of states and, for each $\nu \in \mathcal{P}_2(\mathbb{R}^d)$, let $Q(\nu) = \big(Q(\nu)^{(i,j)}; i, j \in \mathbf{R}\big)$ be the transition semigroup of a continuous-time Markov chain such that each entry $Q^{(i,j)}$ is a Lipschitz mapping on $\mathcal{P}_2(\mathbb{R}^d)$ with respect to the metric $\mathcal{W}_2$. Furthermore, assume that

$$0 < \sup_{\nu \in \mathcal{P}_2(\mathbb{R}^d)} \sup_{i \in \mathbf{R}} \sum_{j \in \mathbf{R}} Q(\nu)^{(i,j)} < \infty.$$

Let $b : \mathcal{P}_2(\mathbb{R}^d) \times \mathbb{R}^d \times \mathbf{R} \to \mathbb{R}^d$ be a bounded function such that $(\nu, x) \mapsto b(\nu, x, e)$ satisfies the corresponding Lipschitz conditions for all e; let σ be a real constant, and $\gamma \equiv 0$. Then, from Theorem 4.1, there exists a solution $\{X^{i,n}, 1 \le i \le n\}$ to the system

$$X_t^{i,n} = X_0^i + \int_0^t b\left(\overline{\mu}_s^n, X_s^{i,n}, Y_s^n\right) \mathrm{d}s + \sigma W_t^i, \qquad \forall i = 1, \ldots, n, \tag{4.32}$$

where $\overline{\mu}^n$ is the empirical measure process from (4.21), and $Y^n = \big\{Y_t^n, t \in [0, T]\big\}$ is the regime

$$Y_t^n := \int_{\{t\} \times \mathbf{R}} r\, \eta^n(\mathrm{d}s, \mathrm{d}r),$$

determined by the intensity kernel

$$K^n(t, \mathrm{d}r) \mathbf{1}_{\{Y_{t-}^n = i\}} = \sum_{j \in \mathbf{R}} Q^{(i,j)}(\overline{\mu}_{t-}^n)\, \delta_{\{j\}}(\mathrm{d}r) = Q^{(i,\cdot)}(\overline{\mu}_{t-}^n) \delta_{\mathbf{R}}(\mathrm{d}r) \quad \forall i \in \mathbf{R}.$$

In other words, $\{X^{i,n}, 1 \le i \le n\}$ above constitutes a state-dependent regime-switching diffusion with transition matrix $Q(\overline{\mu}^n)$ depending on the average position of n particles of the diffusion system, see [70, 71, 77].

Taking a subset $I \subseteq \{1, \ldots, n\}$ fixed, from Theorem 4.2 we then have that $\{X^{i,n}, i \in I, i \le n\}$ converges to an exchangeable collection of conditional McKean-Vlasov regime-switching diffusions $\{X^i, i \in I\}$ as $n \to \infty$. Each X^i is thus a conditionally independent, identically distributed copy of

$$X_t = X_0 + \int_0^t b\left(\mathbb{P}^1\,[X]_s\,, X_s, Y_s\right) \mathrm{d}s + \sigma W_t, \tag{4.33}$$

subjected to the regime $Y = \left\{Y_t := \xi_{N_t^\lambda}, t \in [0, T]\right\}$, where (N^λ, ξ) is the admissible common Poissonian noise with the intensity kernel defined as

$$K^\lambda(t, \mathrm{d}r)\mathbf{1}_{\{Y_{t-}=i\}} = Q^{(i,\cdot)}(\mathbb{P}^1[X]_{t-})\delta_{\mathbf{R}}(\mathrm{d}r), \qquad \forall i \in \mathbf{R} \tag{4.34}$$

Indeed, since each

$$X_0^i + \int_0^t b\left(\overline{\mu}_s^n, X_s^{i,n}, r\right) \mathrm{d}s + \sigma W_t^i$$

converges weakly to

$$X_0^i + \int_0^t b\left(\mathbb{P}^1\left[X^i\right]_s, X_s^i, r\right) \mathrm{d}s + \sigma W_t^i$$

for every $r \in \mathbf{R}$, by introducing the intermediate coupling

$$\underline{X}_t^{i,n} = X_0^i + \int_0^t b\left(\overline{\mu}_s^n, X_s^{i,n}, Y_s\right) \mathrm{d}s + \sigma W_t^i$$

and conditioning with respect to the process N, the well posedness of (4.33) is guaranteed by Theorem 4.2 .

Observe that these results on the conditional propagation of chaos are consistent with those obtained by Shao and Wei [71]. Moreover, they also allow for a direct interaction between the state of the particle system and the regime itself. In terms of optimal control and applications, the previous example describes a game-theoretic parallel to the example of portfolio selection in Sect. 3.5. Indeed, while the situation presented before concerns only one agent aiming to minimize the cost of their investment in an uncertain (random, possibly exogenous) environment, the current example concerns the interaction between many agents. Moreover, it captures the fluctuations in the economy (through the regime) due to the actions of the participants (an endogenous random environment, reflected in the market). With this in mind, a natural question that arises is whether and how these two notions—*optimal control in a random environment* and *propagation of chaos of coupled particles*—can interact with each other. This is one of the motivations behind *mean-field games*, which we address in the following section.

Remark 4.6 One criticism often leveled at the propagation of chaos presented here is the implicit assumption that the coupling between particles is to some extent symmetric. More precisely, the contribution that each particle $X^{i,n}$ provides to the empirical environment $\overline{\mu}^n$ is the same throughout the population. At the time of writing there have been recent developments towards the study of heterogeneous particle systems (e.g. [8, 31, 48], just to name a few); however, their treatment is considerably more involved and falls beyond the scope of our current setting.

4.5 Mean-Field Games with Common Poissonian Noise

Before presenting the main results, we will briefly review the heuristics underlying a mean-field game. First, consider a population of size n in which each of the agents $(i = 1, \dots, n)$ has the task of controlling a stochastic dynamical system of the form

$$X_t^{i,n} = X_0^{i,n} + \int_0^t b\left(\overline{\mu}_{s-}^n, X_{s-}^{i,n}, \alpha_{s-}^{i,n}\right) \mathrm{d}s + \int_0^t \sigma\left(\overline{\mu}_{s-}^n, X_{s-}^{i,n}, \alpha_{s-}^{i,n}\right) \mathrm{d}W_s^i$$
$$+ \int_{(0,t]\times\mathbf{R}} \gamma\left(r, \overline{\mu}_{s-}^n, X_{s-}^{i,n}, \alpha_{s-}^{i,n}\right) \tilde{\eta}^{\lambda,n}(\mathrm{d}s, \mathrm{d}r),$$

where $\overline{\mu}_t^n$ is the empirical measure of the vector $X_t^n = (X_t^{1,n}, \dots, X_t^{n,n})$, $\tilde{\eta}^{\lambda,n}$ is the compensated measure for the lifting $\eta^{\lambda,n}$ given by the empirical noise $(N^{\lambda,n}, \xi^{\lambda,n})$ with kernel $K_t^{\lambda,n}(\mathrm{d}r) = \lambda(r, \overline{\mu}_{t-}^n)Q(\mathrm{d}r)$. Furthermore, each of them aims to minimize its cost functional that is dependent on the state of some statistic of the population X^n:

$$J^{i,n} = J(\overline{\mu}^n, X^{i,n}, \alpha^{i,n}) = \mathbb{E}\left[\int_0^T f(\overline{\mu}_{t-}^n, X_{t-}^{i,n}, \alpha_{t-}^{i,n})\,\mathrm{d}t + g(\overline{\mu}_{T-}^n, X_{T-}^{i,n})\right],$$

for all $i = 1, \dots n$.

Due to the symmetries present in the game, the intuition behind mean-field games consists of approximate the Nash equilibrium of the game described above by analyzing the thermodynamical limit of X^n as $n \to \infty$. Specifically, the idea is to take advantage of the decoupling phenomenon obtained from the propagation of chaos and study the optimal control problem from a representative player in which the state is replaced by a McKean–Vlasov SDE.

For a more detailed explanation of the results presented here, interested readers are referred to [40]. In order to be more precise about the meaning of equilibrium in this context, we recall the notion of an optimal control problem for a generic (fixed) random environment μ, studied in previous chapters:

Let A be a fixed non-empty and compact subset of a finite-dimensional Euclidean space (set of admissible actions). Given the admissible set-up $(X_0, \mu, \lambda, (N^\lambda, \xi), W)$, our aim is to

$$\underset{\substack{\alpha:\Omega\times[0,T]\to A\\ \mathbb{F}-\text{adapted}}}{\text{minimize}} \quad J^{\mu}(X,\alpha), \quad \text{with } J^{\mu}(X,\alpha)$$

$$:= \mathbb{E}\left[\int_0^T f_t^{\mu}(X,\alpha)\,\mathrm{d}t + g(\mu_{T-}, X_{T-})\right], \tag{4.35}$$

subject to the α-controlled dynamics

$$X_t = X_0 + \int_0^t b_s^{\mu}(X,\alpha)\,\mathrm{d}s + \int_0^t \sigma_s^{\mu}(X,\alpha)\,\mathrm{d}W_s + \int_{(0,t]\times\mathbf{R}} \gamma_s^{\mu}(r,X,\alpha)\tilde{\eta}^{\lambda}(\mathrm{d}s,\mathrm{d}r), \tag{4.36}$$

where b, f, g, σ and γ are specified coefficients.

Remark 4.7 Expanding on the observation done in Remark 3.1, observe that by taking μ to be the conditional law of X given a common Poissonian point process, the terminal condition in (4.35) is effectively matched with $\mathbb{E}[g(\mu_T, X_T)]$ since the jump times of both μ and X coincide (Lemma 4.1).

Then, we proceed to define the mean-field game as follows:

Definition 4.2 The *Mean-Field Game with common Poissonian noise* is defined as the problem consisting on finding an $\mathbb{F}$-adapted, *càdlàg* process $(\widehat{\mu}, \widehat{X}) = (\widehat{\mu}, \widehat{X^{\widehat{\mu}}})$, defined on the set-up $(X_0, \widehat{\mu}, \lambda, (N^{\lambda}, \xi), W)$, such that $(\widehat{X}, \widehat{\alpha})$ is an optimal pair for the control problem in random environment (4.35)–(4.36), where $\widehat{\mu}$ is a fixed point of the mapping

$$\mu \longmapsto \mathbb{P}\left[\,\cdot\,\middle|\sigma\{(X_0,\mu,\eta^{\lambda})\}\right] \circ (\widehat{X}^{\mu})^{-1} \equiv \mathbb{P}^1[\widehat{X}^{\mu}]. \tag{4.37}$$

We then refer to $(\widehat{\mu}, \widehat{X})$ as the *Mean-Field Equilibrium (MFE) for the game.*

Now that the game and its equilibrium have been defined, our goal is to explore a way to solve it or, at least, find a set of conditions that guarantee the existence of a solution. Before doing so, we present a brief digression on the nature of the game and its set-up. In fact, a significant detail that needs to be pointed out coming from the previous definition is that, compared to the optimal control problem described in (4.35)–(4.36) where the environment μ is (possibly) taken *exogenously* from the rest of the elements in the set-up $(X_0, \mu, \lambda, (N^{\lambda}, \xi), W)$, in the mean-field game there exists a direct relationship between its solution and the set-up.

To be more precise, the interaction of the environment in the jumps (and thus in the solutions to the control problem) happens through the compensator, specifically through the intensity, while the interaction of the solution in the set-up is explicitly given by the fixed point (4.37). This procedure for describing a mean-field equilibrium $(\widehat{\mu}, \widehat{X})$ makes it clear that the set-up is part of the solution, as it has influence in the components λ and (N^{λ}, ξ), through the random environment $\widehat{\mu}$, which at the same time depends on the conditional law of $\widehat{X}$. This key observation sheds some light on the following aspects:

1. On the role of the compensator as a moderator of the elements involved in the model and, therefore, on the importance of correctly specifying the intensity process.
2. On how the nature of the environment affects the problem in general. In a given *exogenous* environment, the set-up is a priori fixed and solutions to the optimal control problem (4.35)–(4.36) are specified within this fixed setting, while on an *endogenous* environment—as in the case of the mean-field game—the feedback between elements makes it so that the set-up itself is determined by the solution to the control problem, and vice-versa (in fact, this relation environment–solution can be appreciated more clearly later, as we deal with the solutions to the adjoint BSDEs).
3. Why this notion of admissibility (in terms of intensity and set-up) was chosen. Based on the two previous points, an admissible set-up intuitively corresponds to a structure that is rich enough to be able to construct a sequence of iterative optimal control problems (indexed in the space of probability measures), while preserving the underlying Wiener-Poisson structure of the driving noise processes.

4.5.1 Characterization of Mean-Field Equilibria

Let $(\mathbb{P}^1[X^\alpha], X^\alpha)$ be the solution to a conditional McKean-Vlasov equation given the information of the common noise process (N^λ, ξ),

$$\begin{aligned} X_t =& X_0 + \int_0^t b_s^{\mathbb{P}^1[X]}(X,\alpha)\,\mathrm{d}s + \int_0^t \sigma_s^{\mathbb{P}^1[X]}(X,\alpha)\,\mathrm{d}W_s \\ &+ \int_{(0,t]\times\mathbf{R}} \gamma_s^{\mathbb{P}^1[X]}(r,X,\alpha)\tilde{\eta}^\lambda(\mathrm{d}s,\mathrm{d}r), \end{aligned} \tag{4.38}$$

where $\alpha \in \mathcal{A}$ is a given (arbitrary) admissible control, see (C.9) for the definition of admissible strategies. Notice that in this case the conditional law $\mathbb{P}^1[X^\alpha]$ of the process X^α plays the role of the random environment, which in this case is endogenous. This last statement is formalized in Lemma 4.3 below, which acts as a controlled version of Theorem 4.1.

The next results require specific assumptions on the elements of the model, which are selected from the sets of hypotheses introduced in the previous chapters. Below is a brief explanation of their relevance.

Assumptions F

- *(E.1)–(E.2): **Extended assumptions on the space**, page 98.*
- *(D.3): **On the space: Intensity candidate**, page 93.*
- *(D.6): **On the coefficients: Jump condition**, page 93.*
- *(C.4)–(C.6): **On the coefficients**, page 61.*
- *(C.7)–(C.9): **On the controls**, page 61.*

Lemma 4.3 *Assume that Assumptions F hold. Then, Eq.* (4.38) *has a unique strong solution* $(\mathbb{P}^1[X^\alpha], X^\alpha)$ *for any admissible* α*, and consequently, the conditional law* $\mathbb{P}^1[X^\alpha]$ *forms a random environment.*

Remark 4.8 The set of assumptions in the previous lemma allow us to find the disintegration of the base measure $\mathbb{P}$ with respect to the information provided by the common noise, so that the conditional laws $\mathbb{P}^1[X_t]$ are *regular conditional probability measures*, see Sect. 4.2. This fact is then crucial to guarantee that (1) conditional laws are admissible as random environments—Lemma 4.1—, and (2) propagation of chaos results hold, i.e. the convergence of weakly coupled mean-field particle systems to systems with conditional McKean-Vlasov Dynamics as in (4.38)—Theorem 4.2. Moreover: continuing from the discussion presented at the tail end of Sect. 3.2 (definition of the Hamiltonian system), from the information structure presented in (E.1), (E.2) and (D.3), we have that $\mathcal{H}^{2,\perp}$—i.e. the space of martingales that cannot be represented as integrals with respect to neither W nor $\tilde{\eta}^\lambda$—consists of $\{0\}$, as as no additional information outside the system is considered. Consequently, we have that the additional orthogonal martingale terms M and $\widetilde{M}$ in the first and second order adjoint BSDEs, respectively, equal zero.

Proof of Lemma 4.3 For any α admissible and any $x \in \mathcal{S}^2(\mathbb{R}^d)$, define the mapping $\Phi^\alpha : \mathcal{S}^2(\mathbb{R}^d) \to \mathcal{S}^2(\mathbb{R}^d)$ as

$$\Phi^\alpha(x)_t := X_0 + \int_0^t b_s\left(\mathbb{P}^1[x]_{s-}, x_s, \alpha_s\right) \mathrm{d}s + \int_0^t \sigma_s\left(\mathbb{P}^1[x]_{s-}, x_s, \alpha_s\right) \mathrm{d}W_s$$
$$+ \int_{(0,t]\times\mathbf{R}} \gamma_s\left(\mathbb{P}^1[x]_{s-}, r, x_s, \alpha_{s-}\right) \tilde{\eta}^\lambda(\mathrm{d}s, \mathrm{d}r).$$

By following *verbatim* the computations done in Theorem 4.1—i.e. using standard SDE techniques along with the Lipschitz bounds for the integrands—, we obtain that Φ^α is a contractive mapping; consequently, it admits a fixed point. Then, from Corollary 4.1, $\left(X_0, \mathbb{P}^1[X^\alpha], \lambda, (N^\lambda, \xi), W\right)$ is an admissible set-up. This completes the proof. □

With this lemma, we can reformulate the results from previous chapters in the context of mean-field games. Indeed: observe that for an arbitrary control α, the resulting Hamiltonian system is a decoupled FBSDE in the sense that, given the α-controlled system described by (4.38), the corresponding first and second adjoint processes (3.25) and (3.27), respectively, have a known *fixed* driver in the random environment $\mu^\alpha = \mathbb{P}^1[X^\alpha]$. Then, from Theorems 3.1 and 3.2, the optimal control is characterized next; this is the main contribution of this chapter.

Theorem 4.3 *Assume that Assumptions F hold.*

(a) Necessary conditions for equilibrium. *Let* $(\widehat{\mu}, \widehat{X})$ *be a MFE with a corresponding admissible control* $\widehat{\alpha} \in \mathcal{A}$*, such that*

$$\widehat{X}_t \in L^8(\mathbb{P}) \text{ uniformly on } t \in [0, T]. \tag{4.39}$$

Then, for $\mathrm{d}\,\mathbb{P} \otimes \mathrm{d}\,t$*-a.a* $(\omega, t) \in \Omega \times [0, T]$ *there exists a process* $(\widehat{X}, \widehat{Y}, \widehat{Z}, \widehat{U}, \widehat{M})$*, defined as in* (3.31) *with* $\widehat{M} \equiv 0$*, that solves the Conditional McKean-Vlasov FBSDE*

$$\widehat{X}_t = \widehat{X}_0 + \int_0^t \widehat{b}_s \,\mathrm{d}\,s + \int_0^t \widehat{\sigma}_s \,\mathrm{d}\,W_s + \int_{(0,t]\times\mathbf{R}} \widehat{\gamma}_s(r)\tilde{\eta}^\lambda(\mathrm{d}\,s, \mathrm{d}\,r), \tag{4.40}$$

$$\begin{aligned} Y_t =& \mathrm{D}\,\widehat{g}_T^\top + \int_t^T \mathrm{D}\,\widehat{H}_s(Y, Z, U)^\top \,\mathrm{d}\,s - \int_t^T Z_s \,\mathrm{d}\,W_s \\ &- \int_{(t,T]\times\mathbf{R}} U_s(r)\tilde{\eta}^\lambda(\mathrm{d}\,s, \mathrm{d}\,r), \end{aligned} \tag{4.41}$$

$$\begin{aligned} \widetilde{Y}_t =& \mathrm{D}^2\,\widehat{g}_T^\mu + \int_t^T \widehat{B}_s^{\star\star} \,\mathrm{d}\,s - \sum_{i=1}^{n_1} \int_t^T \left\{ \widetilde{Z}_s^{(i)} \,\mathrm{d}\,W_s^{(i)} - (\widehat{\Sigma}_s^{\star\star})^{(i)} \,\mathrm{d}\,s \right\} \\ &- \sum_{j=1}^{n_0} \int_{(t,T]\times\mathbf{R}} \Big\{ \widetilde{U}_s(r)^{(j)}\; \tilde{\eta}^\lambda(\mathrm{d}\,s, \mathrm{d}\,r)^{(j)} \\ &- \widehat{\Gamma}_s^{\star\star}(r)^{(j)} \left(\mathrm{d}\,s \otimes K_s^{\lambda,\widehat{\mu}}(\mathrm{d}\,r)^{(j)} \right) \Big\}, \end{aligned} \tag{4.42}$$

where we have used the notation presented in the previous chapter—see equation (3.3)*—, and* $\widehat{B}^{\star\star}$, $\widehat{\Sigma}^{\star\star}$, $\widehat{\Gamma}^{\star\star}$ *are defined as in* (3.28)–(3.30)*, with* $\mu = \widehat{\mu}$*. In other words,* $(\widehat{X}, \widehat{Y}, \widehat{Z}, \widehat{U}, \widehat{M})$ *is the Hamiltonian system for the optimization problem* (4.35)–(4.36)*, under the endogenous random environment* $\widehat{\mu} = \mathbb{P}^1[\widehat{X}]$.

(b) ***Sufficient conditions for equilibrium.*** *Suppose that the following conditions hold:*

- g *is a convex function on* x*;*
- *the mapping* $(\nu, x, y, z, u) \longmapsto \widehat{a} \in \arg\min_{a\in A} H_t^\nu(x, y, z, u, a)$ *exists* $\mathrm{d}\,\mathbb{P} \otimes \mathrm{d}\,t$*-almost everywhere;*
- *there exists a process* $(\widehat{X}, \widehat{Y}, \widehat{Z}, \widehat{U}, \widehat{M})$ *that solves the conditional McKean-Vlasov FBSDE* (4.40)–(4.42) *for the (admissible) control* $\widehat{\alpha}_t := \arg\min_{a\in A} H_t^{\mathbb{P}^1[\widehat{X}]}\big(\widehat{X}, Y, Z, U, a\big)$*;*
- $\widehat{X}$ *satisfy the moment condition* (4.39)*;*
- *the mapping* $x \longmapsto \min_{a\in A} H_t^{\mathbb{P}^1[\widehat{X}]}(x, Y, Z, U, a)$ *exists and is convex* $\mathrm{d}\,\mathbb{P}\otimes \mathrm{d}\,t$*-a.a.*

Then, by construction, $\mathbb{P}^1[\widehat{X}]$ *solves the fixed point problem* (4.37)*, and consequently,* $(\mathbb{P}^1[\widehat{X}], \widehat{X})$ *is a MFE for the game* (4.35)–(4.36).

Remark 4.9

1. Note that the above result implicitly addresses both problems that arise in defining the mean-field game simultaneously, meaning that, on one hand, it provides a set of forward-backward equations that solve the fixed point problem (4.37)

(since it is a conditional McKean-Vlasov equation), and second, by construction, this system of equations characterizes the optimal control at equilibrium.

2. Notice that Condition (C.3), related to the norm $\|\mathbb{P}^1[\widehat{X}]\|_{\mathcal{W}}$ in (3.18), is analogous to condition (4.39) expressed in terms of the moments of the process $\widehat{X}$.

Proof of Theorem 4.3 Let $\mathbb{E}^1$ denote the expectation operator with respect to the regular conditional probability $\mathbb{P}^1 := \mathbb{P}[\,\cdot\,|\mathbb{F}^{(X_0,\mu,\lambda,(N^\lambda,\xi))}]$ (Lemma 4.1). From the properties of Wasserstein distances and the disintegration of measure $\mathbb{P}$ with respect to $\mathbb{P}^1$,

$$\mathbb{E}[\mathcal{W}_2(\mathbb{P}^1[\widehat{X}]_t, \delta_{\{0\}})^8] \leq \mathbb{E}[\mathcal{W}_8(\mathbb{P}^1[\widehat{X}]_t, \delta_{\{0\}})^8] \leq \mathbb{E}[\mathbb{E}^1[|\widehat{X}_t|^8]]$$

for all $t \in [0, T]$; then, from the moment condition at (4.39) the space conditions, specifically condition (C.3), are satisfied, while the rest are due to the coefficients and the controls themselves.

Now, for part **(a)** let $(\widehat{X}, \widehat{\alpha})$ be the optimal pair for the optimization in random environment (4.35)–(4.36), with $\mu = \mathbb{P}^1[\widehat{X}]$—which exists in view of the assumption that $(\widehat{\mu}, \widehat{X})$ is an equilibrium—. The conclusion follows from a direct application of Theorem 3.1.

Conversely, for part **(b)** it follows that, since $(\widehat{X}, \widehat{Y}, \widehat{Z}, \widehat{U}, \widehat{M})$ solves the conditional McKean-Vlasov FBSDE (4.40)–(4.42), from the forward component (4.40) we deduce that $\widehat{\mu} = \mathbb{P}^1[\widehat{X}]$ constitutes the random environment for the admissible set-up $(\widehat{X}_0, \widehat{\mu}, \lambda, K^\lambda, (N^\lambda, \xi), W)$. Then, from Theorem 3.2, $(\widehat{X}, \widehat{\alpha})$ is the optimal pair from Definition 4.2, with $\widehat{\mu} = \mathbb{P}^1[\widehat{X}]$ the solution to the fixed point problem (4.37) by virtue of being part of the solution to the conditional McKean-Vlasov equation (4.40). □

Remark 4.10 When the diffusion and jump terms (σ and γ respectively) do not depend on the control, then the Hamiltonian system $\left(\widehat{X}, \widehat{Y}, \widehat{Z}, \widehat{U}, \widehat{M}\right)$ from Theorem 4.3 is reduced to $\left(\widehat{X}, Y, Z, U, M\right)$, where $\widehat{X} \in \mathcal{S}^2(\mathbb{R}^d)$ satisfies (4.40) and (Y, Z, U, M), defined as in (3.24) with $M = 0$, is a solution to (4.41). That is, $(\widetilde{Y}, \widetilde{Z}, \widetilde{U}, \widetilde{M})$ is identically zero. Furthermore, the extended Pontryagin condition (3.45) is reduced to (3.53):

$$\widehat{H}_t = H_t^{\widehat{\mu}}\left(\widehat{X}, \widehat{Y}, \widehat{Z}, \widehat{U}, \widehat{\alpha}\right) = \operatorname*{ess\,inf}_{\alpha \in \mathcal{A}} H_t^{\widehat{\mu}}\left(\widehat{X}, \widehat{Y}, \widehat{Z}, \widehat{U}, \alpha\right), \qquad \forall t \in [0, T].$$

Observe that the results we have presented so far are consistent with their analogs analyzed in [23, 24] and the references there-in; namely, those regarding the solution of MFGs with a common Gaussian noise.

4.6 The Linear-Quadratic Model, Revisited

To conclude this chapter, we apply the results obtained in the previous sections to characterize the solutions of a *linear-quadratic model* for dynamical systems with self-exciting jumps, as well as the mean-field game derived from it. To be precise, we shall follow the dynamics introduced in Sect. 2.4.3 of Chap. 2 to characterize the solutions to the optimal control problem presented in Sect. 3.1.2 of Chap. 3, together with their counterparts in the mean-field game.

4.6.1 Optimal Control Problem

Let μ be a given random environment and let N^λ be an unmarked Poissonian common noise with intensity λ as in (3.10) from Chap. 3—compare also with (2.52) from Chap. 2—:

$$\lambda_t = m_{k_0}^2(\mu_{t-}) + k_1 + \int_{[0,t)} e^{-k_2(t-s)}\,\mathrm{d}N_s^\lambda, \qquad \forall t > 0, \tag{4.43}$$

where $k_0, k_1, k_2 > 0$ are given parameters and $m_k^2(\nu)$ denotes the second truncated moment of ν, defined in (2.49):

$$m_k^2(\nu) := \int_{\mathbb{R}^d} \left(k\,\frac{x}{\max\{k,\,\|x\|\}}\right)^2 \nu(\mathrm{d}x), \tag{2.49}$$

again, λ is an additive intensity; see Sect. 2.4.1.

Let $A = \mathbb{R}$ be the set of actions, $\mathcal{A}$ the set of admissible controls as in Condition (C.9) (*càdlàg*, $\mathbb{F}$-adapted, $\alpha_t \in L^8(\mathbb{P})$ uniformly in $[0,T]$), and consider the optimal control given by the payoff functional

$$\begin{aligned}\mathbb{E}\Big[\int_0^T \{\tfrac{f_0}{2}(X_{t-}-m(\mu_{t-}))^2 + f_1(X_{t-}-m(\mu_{t-}))\alpha_t + \tfrac{f_2}{2}\alpha_t^2\}\,\mathrm{d}t \\ + \tfrac{g}{2}(X_{T-}-m(\mu_{T-}))^2\Big],\end{aligned} \tag{4.44}$$

where $f_0, f_2, g > 0$ and $f_1 \in \mathbb{R}$; the function m is the mean value function from (2.48):

$$m(\nu) = \int_{\mathbb{R}^d} x\,\nu(\mathrm{d}x); \tag{2.48}$$

and the state X evolves according to the controlled dynamics

$$X_t = X_0 + \int_0^t \{b\,(X_{s-} - m(\mu_{s-})) + \alpha_s\}\,\mathrm{d}s + \sigma W_t + \gamma \tilde{N}_t^{\lambda}, \tag{4.45}$$

for $b \in \mathbb{R}$, $\sigma > 0$ and $\gamma \neq 0$ constants.

It is a known fact that both m is a Lipschitz function w.r.t. the 2-Wasserstein distance; again, see Sect. 2.4.1. Then, from (3.17), the pathwise admissible Hamiltonian on the event $\{\mu_{t-} = \nu, U_t = u\}$ is

$$\begin{aligned} H_t^{\nu}(x, y, z, u, a) :=& \frac{f_0}{2}(x - m(\nu))^2 + f_1(x - m(\nu))a + \frac{f_2}{2}a^2 \\ &+ b\,(x - m(\nu))\,y + ay + \sigma z + \gamma u \lambda_t. \end{aligned} \tag{4.46}$$

Moreover, from the positivity of coefficients in (4.44) we obtain that H_t^{ν} is convex in x (observe that $\mathrm{D}^2 H_t^{\nu} = f_0 > 0$) and it reaches its minimum with respect to a at the point

$$\widehat{a} = -\frac{f_1(x - m(\nu)) + y}{f_2} =: \widehat{\mathbf{a}}(\nu, x, y). \tag{4.47}$$

We now proceed to apply the maximum principle. From Theorem 3.1 (and Remark 3.8 for this particular case), necessary conditions for optimality dictate that if $(\widehat{X}, \widehat{\alpha})$ is an optimal pair for (4.44)–(4.45), then $\widehat{\alpha}_{\cdot} := \widehat{\mathbf{a}}(\mu_{\cdot -}, X_{\cdot -}, Y_{\cdot -})$ where $\widehat{X}$ solves the $\widehat{\alpha}$-controlled SDE (4.45), and $\widehat{Y}$ is the first component of the solution to the BSDE

$$\begin{aligned} \widehat{Y}_t =& g\left(\widehat{X}_{T-} - m\left(\mu_{T-}\right)\right) + \int_t^T \left\{f_0\left(\widehat{X}_{s-} - m\left(\mu_{s-}\right)\right) + b\widehat{Y}_{s-} + f_1\widehat{\alpha}_s\right\}\mathrm{d}s \\ &- \int_t^T \widehat{Z}_s\,\mathrm{d}W_s - \int_t^T \widehat{U}_s\,\mathrm{d}\tilde{N}_s^{\lambda} - (\widehat{M}_T - \widehat{M}_t). \end{aligned}$$

Conversely, let $\widehat{\mathbf{a}}(\nu, x, y) = \widehat{a}$ be the candidate policy for optimality obtained by minimizing the Hamiltonian (4.47). If the coupled FBSDE

$$X_t = X_0 + \int_0^T \{b\,(X_{s-} - m(\mu_{s-})) + \widehat{\mathbf{a}}(\mu_{s-}, X_{s-}, Y_{s-})\}\,\mathrm{d}s + \sigma W_t + \gamma \tilde{N}_t^{\lambda}, \tag{4.48}$$

$$\begin{aligned} Y_t =& g(X_{T-} - \mu_{T-}) + \int_t^T \{f_0\,(X_{s-} - m\,(\mu_{s-})) + bY_{s-} + f_1\widehat{\mathbf{a}}(\mu_{s-}, X_{s-}, Y_{s-})\}\,\mathrm{d}s \\ &- \int_t^T Z_s\,\mathrm{d}W_s - \int_t^T U_s\,\mathrm{d}\tilde{N}_s^{\lambda} - (M_T - M_t), \end{aligned} \tag{4.49}$$

has a solution (X, Y, Z, U, M), then, from Theorem 3.2, $(X, \widehat{\mathbf{a}}(\mu_{\cdot-}, X_{\cdot-}, Y_{\cdot-}))$ is an optimal pair for (4.44)–(4.45), where the admissiblity of the control is due to the linearity of the coefficients and the a priori estimates of the states from Lemma 3.1.

Remark 4.11 Regarding the solutions to (4.48)–(4.49), observe that the FBSDE can be rewritten as the coupled system of linear equations

$$X_t = X_0 + \int_0^t \{B\,(X_{s-} - m(\mu_{s-})) - F_2 Y_{s-}\}\,\mathrm{d}s + \sigma W_t + \gamma \tilde{N}_t^{\lambda}, \tag{4.48*}$$

$$\begin{aligned} Y_t =& Y_T + \int_t^T \{F_1(X_{s-} - m\,(\mu_{s-})) + B Y_{s-}\}\,\mathrm{d}s \\ &- \int_t^T Z_s\,\mathrm{d}W_s - \int_t^T U_s\,\mathrm{d}\tilde{N}_s^{\lambda} - (M_T - M_t), \end{aligned} \tag{4.49*}$$

where $Y_T = g(X_{T-} - \mu_{T-})$ and

$$B := b - \frac{f_1}{f_2}, \qquad F_1 := f_0 - \frac{f_1^2}{f_2}, \qquad F_2 := \frac{1}{f_2}. \tag{4.50}$$

Following the results from Theorem 2.1, (4.48*)–(4.49*) has a unique strong solution if there exist non-negative constants $\beta_1, \beta_2, \beta_3$ with $\beta_1 + \beta_2 > 0$ or $\beta_2 + \beta_3 > 0$ such that the following monotonicity conditions hold for every $(\delta x, \delta y, \delta z, \delta u) = (x - x', y - y', z - z', u - u')$:

$$\left(\begin{bmatrix} -F_1\delta x - B\delta y \\ B\delta x - F_2\delta y \\ 0 \\ 0 \end{bmatrix} \cdot \begin{bmatrix} \delta x \\ \delta y \\ \delta z \\ \delta u \end{bmatrix} \right) \leq -\beta^1\,|\delta x|^2 - \beta^2\left[|\delta y|^2 + |\delta z|^2 + \lambda_t\,|\delta u|^2\right]$$

and

$$g\,|\delta x|^2 \geq \beta^3\,|\delta x|^2\,.$$

Owing to the fact that coefficients defining the payoff functional at equation (4.44) are positive, the previous condition is valid whenever $\beta_1 = F_1 > 0$, $\beta_2 = 0$ and $\beta_3 = g$. In terms of the original FBSDE, this means that the Hamiltonian system (4.48)–(4.49) (and by extension, the optimal control problem) has a solution if the coefficients f_0, f_1, f_2, g satisfy the inequalities

$$f_0, f_2, g > 0 \qquad \text{and} \qquad f_0 f_2 > f_1^2.$$

More succinctly: under the conditions presented in Eq. (4.50) and applying Theorem 2.1, the Hamiltonian system for the linear quadratic model is uniquely solvable, provided that μ is a given exogenous environment.

4.6.2 Mean-Field Game

Concerning the mean-field game, we know from Theorem 4.3 that under our prescribed conditions (size of the jumps constant) the parametrized SDE (4.45) with $\mu = \mathbb{P}^1[X^\alpha]$ has a unique solution $(\mathbb{P}^1[X^\alpha], X^\alpha)$ for any admissible control α. Moreover: the corresponding $\mathbb{P}^1[X^\alpha]$ then forms an endogenous random environment; thus, the parameterized BSDE

$$\begin{aligned} Y_t^\alpha =& Y_T + \int_t^T \left\{ f_0 \left(X_{s-}^\alpha - \mathbb{E}^1 \left[X_{s-}^\alpha \right] \right) + bY_{s-}^\alpha + f_1 \alpha_s \right\} \mathrm{d}\, s \\ & - \int_t^T Z_s^\alpha \,\mathrm{d}\, W_s - \int_t^T U_s^\alpha \,\mathrm{d}\, \tilde{N}_s^\lambda \end{aligned}$$

with $Y_T = g\left(X_{T-}^\alpha - \mathbb{E}^1 \left[X_{T-}^\alpha \right] \right)$, has a solution $(Y^\alpha, Z^\alpha, U^\alpha)$; indeed, observe that this type of BSDE falls under the general framework of [58], namely Theorem 3.5 and Corollary 3.6. Then, from Theorem 4.3 and our previous computations, the mean-field game

$$\begin{aligned} \underset{\alpha \in \mathcal{A}}{\text{minimize}} \quad & \Bigg\{ \mathbb{E}\Bigg[\int_0^T \{ \frac{f_0}{2} \left(X_{t-} - \mathbb{E}^1[X_{t-}] \right)^2 \\ & \qquad + f_1 \left(X_{t-} - \mathbb{E}^1[X_{t-}] \right) \alpha_t + \frac{f_2}{2} \alpha_t^2 \} \,\mathrm{d}\, t \Bigg] \\ & \qquad + \mathbb{E}\left[\frac{g}{2} \left(X_{T-} - \mathbb{E}^1[X_{t-}] \right)^2 \right] \Bigg\} \end{aligned}$$

subject to

$$\begin{aligned} X_t =& X_0 + \int_0^t \left\{ b \left(X_{t-} - \mathbb{E}^1 \left[X_{t-} \right] \right) + \alpha_s \right\} \mathrm{d}\, s + \sigma W_t + \gamma \tilde{N}_t^\lambda, \\ \lambda_t =& \mathbb{E}^1 \left[(-k_0 \mathbf{1}_{(-\infty, -k_0)} + X_{t-} \mathbf{1}_{[-k_0, k_0]} + k_0 \mathbf{1}_{(k_0, \infty)})^2 \right] \\ & + k_1 + \int_{[0,t)} e^{-k_2 (t-s)} \,\mathrm{d}\, N_s^\lambda, \end{aligned}$$

has a mean-field equilibrium $(\widehat{\mu}, \widehat{X})$ and optimal control $\widehat{\alpha} \in \mathcal{A}$ if and only if $\widehat{\alpha}_t = \widehat{\mathbf{a}}(\mathbb{P}^1[\widehat{X}]_{t-}, \widehat{X}_{t-}, \widehat{Y}_{t-}) = -\frac{f_1}{f_2}(\widehat{X}_{t-} - \mathbb{E}^1[\widehat{X}_{t-}]) - \frac{1}{f_2}\widehat{Y}_{t-}$ for all t, where $(\widehat{X}, \widehat{Y}, \widehat{Z}, \widehat{U})$ solves the linear McKean-Vlasov FBSDE

$$\widehat{X}_t = X_0 + \int_0^T \left\{ B \left(\widehat{X}_{s-} - \mathbb{E}^1 \left[\widehat{X}_{s-} \right] \right) - F_2 \, \widehat{Y}_{s-} \right\} \mathrm{d}\, s + \sigma W_t + \gamma \tilde{N}_t^\lambda,$$

$$
\begin{aligned}
\widehat{Y}_t =& \widehat{Y}_T + \int_t^T \left\{ F_1 \left(\widehat{X}_{s-} - \mathbb{E}^1 \left[\widehat{X}_{s-} \right] \right) + B\, \widehat{Y}_{s-} \right\} \mathrm{d}\, s \\
& - \int_t^T \widehat{Z}_s \, \mathrm{d}\, W_s - \int_t^T \widehat{U}_s \, \mathrm{d}\, \tilde{N}_s^{\lambda}, \\
\lambda_t =& \mathbb{E}^1 \left[(-k_0 \mathbf{1}_{(-\infty,-k_0)} + \widehat{X}_{t-} \mathbf{1}_{[-k_0,k_0]} + k_0 \mathbf{1}_{(k_0,\infty)})^2 \right] \\
& + k_1 + \int_{[0,t)} e^{-k_2(t-s)} \, \mathrm{d}\, N_s^{\lambda},
\end{aligned}
$$

with B, F_1, F_2 as in (4.50), the terminal condition given by $\widehat{Y}_T = g(\widehat{X}_{T-} - \mathbb{E}^1[\widehat{X}_{T-}])$; indeed, observe that $\mathbb{E}^1[\widehat{X}]$ and $\mathbb{E}^1[\widehat{X}^2]$ are linear functions on the measure argument.

References

1. N. Agram, B. Øksendal, A Hida-Malliavin white noise calculus approach to optimal control. Infinite Dimen. Anal. Quant. Probab. Relat. Topics **21**(03), 1850014 (2018). ISSN: 0219-0257, 1793–6306. https://doi.org/10.1142/S0219025718500145
2. N. Agram, B. Øksendal, Stochastic Fokker-Planck equations for conditional McKean-Vlasov jump diffusions and applications to optimal control. SIAM J. Control Optim. **61**(3), 1472–1493 (2023). ISSN: 0363-0129, 1095-7138. https://doi.org/10.1137/21M1461034
3. A. Aksamit, M. Jeanblanc, *Enlargement of Filtration with Finance in View*. SpringerBriefs in Quantitative Finance (Springer, Cham, 2017). https://doi.org/10.1007/978-3-319-41255-9
4. L. Andreis, P. Dai Pra, M. Fischer, McKean-Vlasov limit for interacting systems with simultaneous jumps. Stoch. Anal. Appl. **36**(6), 960–995 (2018). https://doi.org/10.1080/07362994.2018.1486202
5. V. Bally, L. Caramellino, R. Cont, *Stochastic Integration by Parts and Functional Itô Calculus*, ed. by F. Utzet, J. Vives. Advanced Courses in Mathematics - CRM Barcelona (Springer, Cham, 2016). isbn: 978-3-319-27127-9. https://doi.org/10.1007/978-3-319-27128-6. Accessed 24 April 2026
6. V. Bally, D. Goreac, V. Rabiet, Regularity and stability for the semigroup of jump diffusions with state-dependent intensity. Ann. Appl. Probab. **28**(5), 3028–3074 (2018). ISSN: 1050-5164. https://doi.org/10.1214/18-AAP1382
7. M. Barczy, Z. Li, G. Pap, Yamada-Watanabe results for stochastic differential equations with jumps. Int. J. Stoch. Anal. **2015**, 1–23 (2015). https://doi.org/10.1155/2015/460472
8. E. Bayraktar, S. Chakraborty, R. Wu, Graphon mean field systems. Ann. Appl. Probab. **33**(5), 3587–3619 (2023). ISSN: 1050-5164. https://doi.org/10.1214/22-AAP1901. Accessed 11 April 2024
9. R.E. Bellman, *An Introduction to the Theory of Dynamic Programming* (RAND Corporation, Santa Monica, 1953)
10. R.E. Bellman, Dynamic programming and a new formalism in the calculus of variations. Proc. Natl. Acad. Sci. USA **40**(4), 231–235 (1954). ISSN: 00278424, 10916490
11. R.E. Bellman, On the theory of dynamic programming. Proc. Natl. Acad. Sci. USA **38**(8), 716–719 (1954). ISSN: 00278424, 10916490
12. A. Bensoussan, Stochastic maximum principle for distributed parameter systems. J. Frank. Inst. **315**(5–6), 387–406 (1983). https://doi.org/10.1016/0016-0032(83)90059-5
13. P. Billingsley, *Convergence of Probability Measures* (Wiley, Hoboken, 2013), 247pp. Google Books: 6ItqtwaWZZQC

D. Hernández-Hernández, J. H. Ricalde-Guerrero, *Optimal Control in Random Environments*, SpringerBriefs in Mathematics,
https://doi.org/10.1007/978-3-032-25641-6

14. J.-M. Bismut, Conjugate convex functions in optimal stochastic control. J. Math. Anal. Appl. **44**(2), 384–404 (1973). https://doi.org/10.1016/0022-247X(73)90066-8
15. P. Blanchard, E. Brüning, *Mathematical Methods in Physics: Distributions, Hilbert Space Operators, Variational Methods, and Applications in Quantum Physics*. Progress in Mathematical Physics, vol. 69 (Springer, Cham, 2015). isbn: 978-3-319-14044-5. https://doi.org/10.1007/978-3-319-14045-2. Accessed 24 Feb 2026
16. V.G. Boltyansky, The maximum principle in the theory of optimal processes. Dokl. Akad. Nauk SSSR **119**(6), 1070–1073 (1958)
17. V.G. Boltyansky, R.V. Gamkrelidze, L.S. Pontryagin, On the theory of optimal processes. Dokl. Akad. Nauk SSSR **110**(1), 7–10 (1956)
18. V.G. Boltyansky, R.V. Gamkrelidze, L.S. Pontryagin, Theory of optimal processes. I. The maximum principle. Iz. Akad. Nauk SSSR. Ser. Mat. **24**(1), 3–42 (1960)
19. A. Brausmann, L. Bretschger, Escaping Damocles' sword: en-dogenous climate shocks in a growing economy. Environ. Resour. Econ. **87**(6), 1545–1592 (2024). ISSN: 0924-6460, 1573–1502. https://doi.org/10.1007/s10640-023-00835-w
20. P. Bre'maud, *Point Process Calculus in Time and Space: An Introduction with Applications*. Probability Theory and Stochastic Modelling, vol. 98 (Springer, Cham, 2020). https://doi.org/10.1007/978-3-030-62753-9.
21. A. Cadenillas, I. Karatzas, The Stochastic maximum principle for linear, convex optimal control with random coefficients. SIAM J. Control Optim. **33**(2), 590–624 (1995). ISSN: 0363-0129, 1095-7138. https://doi.org/10.1137/S0363012992240722. Accessed Dec 02 2026
22. R. Carmona, *Applications of Mean Field Games in Financial Engineering and Economic Theory* (2020). arXiv: 2012.05237 [econ, math, q-fin]. https://doi.org/10.48550/arXiv.2012.05237
23. R. Carmona, F. Delarue, *Probabilistic Theory of Mean Field Games with Applications I*. Probability Theory and Stochastic Modelling, vol. 83 (Springer, Cham, 2018). https://doi.org/10.1007/978-3-319-58920-6.
24. R. Carmona, F. Delarue, *Probabilistic Theory of Mean Field Games with Applications II*. Probability Theory and Stochastic Modelling, vol. 84 (Springer, Cham, 2018). https://doi.org/10.1007/978-3-319-56436-4
25. L.-P. Chaintron, A. Diez, *Propagation of Chaos: A Review of Models, Methods and Applications. I. Models and Methods* (2022). https://doi.org/10.48550/arXiv.2203.00446. arXiv: 2203.00446 [math-ph]
26. L.-P. Chaintron, A. Diez, *Propagation of Chaos: A Review of Models, Methods and Applications. II. Applications* (2022). https://doi.org/10.48550/arXiv.2106.14812. arXiv: 2106.14812 [math-ph]
27. F.H. Clarke, *Functional Analysis, Calculus of Variations and Optimal Control*. Graduate Texts in Mathematics, vol. 264 (Springer, London, 2013). isbn: 978-1-4471-4819-7. https://doi.org/10.1007/978-1-4471-4820-3
28. S.N. Cohen, R.J. Elliott, *Stochastic Calculus and Applications*. Probability and Its Applications, 2nd edn. (Birkhäuser, New York, 2015)
29. D.J. Daley, D. Vere-Jones, *An Introduction to the Theory of Point Processes. Volume II: General Theory and Structure*. Red. by J. Gani et al. Probability and Its Applications (Springer, New York, 2008). https://doi.org/10.1007/978-0-387-49835-5
30. D. Dawson, Measure-valued markov processes, in *Ecole d'Eté de Probabilités de Saint-Flour XXI - 1991*, ed. by D.A. Dawson, B. Maisonneuve, J. Spencer, Paul-Louis Hennequin, vol. 1541 (Springer, Berlin, 1993), pp. 1–260. https://doi.org/10.1007/BFb0084190
31. A. De Crescenzo, et al., *Mean-field control of non exchangeable systems*. Control, Optimisation and Calculus of Variations (2024). https://doi.org/10.48550/ARXIV.2407.18635. Accessed 06 Dec 2025
32. F. Delarue, et al., Global solvability of a networked integrate-and-fire model of McKean-Vlasov type. Ann. Appl. Probab. **25**(4), 2096–2133 (2015). ISSN: 1050-5164. https://doi.org/10.1214/14-AAP1044

33. S. Delattre, N. Fournier, M. Hoffmann, Hawkes processes on large networks. Ann. Appl. Probab. **26**(1), 216–261 (2016). https://doi.org/10.1214/14-AAP1089
34. N. El Karoui, S. Peng, M.C. Quenez, Backward stochastic differential equations in finance. Math. Financ. **7**(1), 1–71 (1997). https://doi.org/10.1111/1467-9965.00022
35. N.C. Framstad, B. Øksendal, A. Sulem, Sufficient stochastic maximum principle for the optimal control of jump diffusions and applications to finance. J. Optim. Theory Appl. **121**(1), 77–98 (2004). https://doi.org/10.1023/B:JOTA.0000026132.62934.96
36. R.V. Gamkrelidze, On the general theory of optimal processes. Dokl. Akad. Nauk SSSR **123**(2), 223–226 (1958)
37. I.M. Gel'fand, et al., *Calculus of Variations*, ed. by R.A. Silverman. Unabridged reprint (Dover Publications, Mineola, 2000). isbn: 978-0-486-41448-5
38. D. Hernández-Hernández, J.H. Ricalde-Guerrero, Conditional McKean-Vlasov differential equations with common poissonian noise: propagation of chaos. Stoch. Dyn. **25**, 2550033 (2025). ISSN: 0219-4937, 1793-6799. https://doi.org/10.1142/S0219493725500339
39. D. Hernández-Hernández, J.H. Ricalde-Guerrero, Coupled forward-backward stochastic differential equations with jumps in random environments. Stoch. Anal. Appl. **44**, 35–64 (2025). ISSN: 0736-2994, 1532-9356. https://doi.org/10.1080/07362994.2025.2594452
40. D. Hernández-Hernández, J.H. Ricalde-Guerrero, Mean-field games with common Poissonian noise: a maximum principle approach. Math. Control Relat. Fields (2026). ISSN: 2156-8472, 2156–8499. https://doi.org/10.3934/mcrf.2026015
41. N. Ikeda, S. Watanabe, *Stochastic Differential Equations and Diffusion Processes*, 2nd edn. (North-Holland, Amsterdam, 1989)
42. J. Jacod, *Calcul Stochastique et Proble'mes de Martingales* (Springer, Berlin, 2006), 550pp. Google Books: 2Dd6CwAAQBAJ
43. J. Jacod, A.N. Shiryaev, *Limit Theorems for Stochastic Processes*, Red. by A. Chenciner et al. Grundlehren Der Mathematischen Wissenschaften, vol. 288 (Springer, Berlin, 2003). https://doi.org/10.1007/978-3-662-05265-5
44. M. Jóźwikowski, W. Respondek, A contact covariant approach to optimal control with applications to sub-riemannian geometry. Math. Control Signals Syst. **28**(3), 27 (2016). ISSN: 0932-4194, 1435-568X. https://doi.org/10.1007/s00498-016-0176-3
45. J.F.C. Kingman, *Poisson Processes*. Oxford Studies in Probability, vol. 3 (Clarendon Press, Oxford, 1993). 104pp.
46. J.F.C. Kingman, Poisson processes revisited. Probab. Math. Stat. **26**(1), 75–95 (2006)
47. H.J. Kushner, Necessary conditions for continuous parameter stochastic optimization problems. SIAM J. Control **10**(3), 550–565 (1972). https://doi.org/10.1137/0310041
48. D. Lacker, A. Soret, A label-state formulation of stochastic graphon games and approximate equilibria on large networks. Math. Oper. Res. **48**, 1811–2382 (2023). moor.2022.1329. ISSN: 0364-765X, 1526-5471. https://doi.org/10.1287/moor.2022.1329. Accessed 11 Feb 2026
49. T. Li, et al., A review of dynamic Stackelberg game model. Discrete Contin. Dyn. Syst. B **22**(1), 125–159 (2017). ISSN: 1553-524X. https://doi.org/10.3934/dcdsb.2017007
50. X. Li, Y. Yao, Maximum principle of distributed parameter systems with time lags, in *Distributed Parameter Systems*, ed. by F. Kappel, K. Kunisch, W. Schappacher, vol. 75 (Springer, Berlin, 1985), pp. 410–427. https://doi.org/10.1007/BFb0005665
51. E. Löcherbach, Convergence to equilibrium for time-inhomogeneous jump diffusions with state-dependent jump intensity. J. Theor. Probab. **33**(4), 2280–2314 (2020). ISSN: 0894-9840, 1572-9230. https://doi.org/10.1007/s10959-019-00947-4
52. X. Mao, *Stochastic Differential Equations and Applications*, 2nd edn. (Woodhead Publishing, Cambridge, 2011)
53. S. Mehri, et al., Propagation of chaos for stochastic spatially structured neuronal networks with delay driven by jump diffusions. Ann. Appl. Probab. **30**(1), 175–207 (2020). ISSN: 1050-5164. https://doi.org/10.1214/19-AAP1499
54. S.L. Nguyen, G. Yin, D.T. Nguyen, A general stochastic maximum principle for mean-field controls with regime switching. Appl. Math. Optim. **84**(3), 3255–3294 (2021). https://doi.org/10.1007/s00245-021-09747-x

55. W. Oettli, D. Schläger, Existence of equilibria for g-monotone mappings. IFAC Proc. Vols **31**(13), 13–17 (1998). https://doi.org/10.1016/S1474-6670(17)35990-6
56. W. Oettli, D. Schläger, Generalized vectorial equilibria and generalized monotonicity, in *Functional Analysis with Current Applications in Science, Technology and Industry*, ed. by M. Brokate, A.H. Siddiqi (Chapman and Hall/CRC, New York, 2021). https://doi.org/10.1201/9780429332913
57. B. Øksendal, A. Sulem, *Applied Stochastic Control of Jump Diffusions*. Universitext (Springer, Cham, 2019). https://doi.org/10.1007/978-3-030-02781-0
58. A. Papapantoleon, D. Possamaï, A. Saplaouras, Existence and uniqueness results for BSDE with jumps: the whole nine yards. Electron. J. Probab. **23**, 1–68 (2018). https://doi.org/10.1214/18-EJP240
59. E. Pardoux, S.G. Peng, Adapted solution of a backward stochastic differential equation. Syst. Control Lett. **14**(1), 55–61 (1990). https://doi.org/10.1016/0167-6911(90)90082-6
60. G. Peccati, M. Reitzner (Eds.), *Stochastic Analysis for Poisson Point Processes: Malliavin Calculus, Wiener-Itô Chaos Expansions and Stochastic Geometry*. Bocconi & Springer Series, Mathematics, Statistics, Finance and Economics, vol. 7 (Springer, Cham, 2016). isbn: 978-3-319-05233-5. https://doi.org/10.1007/978-3-319-05233-5
61. S. Peng, A general stochastic maximum principle for optimal control problems. SIAM J. Control Optim. **28**(4), 966–979 (1990). https://doi.org/10.1137/0328054
62. S. Peng, Z. Wu, Fully coupled forward-backward stochastic differential equations and applications to optimal control. SIAM J. Control Optim. **37**(3), 825–843 (1999). https://doi.org/10.1137/S0363012996313549
63. L.S. Pontryagin, Optimal control processes. Uspekhi Mat. Nauk **14**, 3–20 (1959)
64. P.D. Pra, F.D. Hollander, McKean-Vlasov limit for interacting random processes in random media. J. Stat. Phys. **84**(3–4), 735–772 (1996). ISSN: 0022-4715, 1572-9613. https://doi.org/10.1007/BF02179656
65. P.E. Protter, *Stochastic Integration and Differential Equations*. Stochastic Modelling and Applied Probability, vol. 21 (Springer, Berlin, 2005). https://doi.org/10.1007/978-3-662-10061-5
66. M.-C. Quenez, A. Sulem, BSDEs with jumps, optimization and applications to dynamic risk measures. Stoch. Process. Their Appl. **123**(8), 3328–3357 (2013). https://doi.org/10.1016/j.spa.2013.02.016
67. S.I. Resnick, *A Probability Path*. Modern Birkhäuser Classics (Birkhäuser, Boston, 2014). isbn: 978-0-8176-8409-9. https://doi.org/10.1007/978-0-8176-8409-9. Accessed 21 Aug 2025
68. D. Sakhanda, J.H. Ricalde-Guerrero, *Infinite-Horizon Optimal Control of Jump-Diffusion Models for Pollution-Dependent Disasters* (2025). arXiv:2511.13568 [math]. https://doi.org/10.48550/arXiv.2511.13568
69. F. Santambrogio, *Optimal Transport for Applied Mathematicians*, Progress in Nonlinear Differential Equations and Their Applications, vol. 87 (Springer, Cham, 2015). https://doi.org/10.1007/978-3-319-20828-2
70. J. Shao, Strong solutions and strong feller properties for regime-switching diffusion processes in an infinite state space. SIAM J. Control Optim. **53**(4), 2462–2479 (2015). https://doi.org/10.1137/15M1013584
71. J. Shao, D. Wei, Propagation of chaos and conditional McKean-Vlasov SDEs with regime-switching. Front. Math. China **17**(4), 731–746 (2022). https://doi.org/10.1007/s11464-021-0960-3
72. A.-S. Sznitman, Topics in propagation of chaos, in *Ecole d'Eté de Probabilités de Saint-Flour XIX - 1989*, ed. by D.L. Burkholder, E. Pardoux, A.-S. Sznitman, P.-L. Hennequin, vol. 1464 (Springer, Berlin, 1991), pp. 165–251. https://doi.org/10.1007/BFb0085169
73. S. Tang, X. Li, Necessary conditions for optimal control of stochastic systems with random jumps. SIAM J. Control Optim. **32**(5), 1447–1475 (1994). https://doi.org/10.1137/S0363012992233858

74. N. Touzi, A. Tourin, *Optimal Stochastic Control, Stochastic Target Problems, and Backward SDE*. Fields Institute Monographs, vol. 29 (Fields Institute for Research in the Mathematical Sciences Springer, New York, 2013). isbn: 978-1-4614-4286-8
75. Z. Xun yu, A unified treatment of maximum principle and dynamic programming in stochastic controls. Stoch. Stoch. Rep. **36**(3–4), 137–161 (1991). https://doi.org/10.1080/17442509108833715
76. T. Yamada, S. Watanabe, On the uniqueness of solutions of stochastic differential equations. Kyoto J. Math. **11**(1), 155–167 (1971). https://doi.org/10.1215/kjm/1250523691
77. G.G. Yin, C. Zhu, *Hybrid Switching Diffusions*. Stochastic Modelling and Applied Probability, vol. 63 (Springer, New York, 2010). https://doi.org/10.1007/978-1-4419-1105-6
78. J. Yong, X.Y. Zhou, *Stochastic Controls: Hamiltonian Systems and HJB Equations*. Applications of Mathematics, vol. 43 (Springer, New York, 1999), 438pp.
79. W. Zhen, Forward-backward stochastic differential equations with brownian motion and poisson process. Acta Math. Appl. Sin. **15**(4), 433–443 (1999). https://doi.org/10.1007/BF02684045

Index

D. Hernández-Hernández, J. H. Ricalde-Guerrero, *Optimal Control in Random Environments*, SpringerBriefs in Mathematics,
https://doi.org/10.1007/978-3-032-25641-6

The manufacturer's authorised representative in the EU is Springer Nature Customer Service Centre GmbH, Europaplatz 3, 69115 Heidelberg, Germany. If you have any concerns regarding our products, please contact ProductSafety@springernature.com

Printed and bound by CPI Group (UK) Ltd, Croydon, CR0 4YY
10/08/2026
02192373-0001